Miller, L. G. and Murray, W. J. (Eds.). (1998). *Herbal Medicinals: A Clinician's Guide,* The Haworth Press, Inc., Binghamton, New York.

Mutschler, E., Geisslinger, G., Kroemer, H. K., and Schäfer-Korting, M. (2001). *Mutschler Arzneimittelwirkungen, Lehrbuch der Pharmakologie und Toxikologie,* Wissenschaftliche Verlagsgesellschaft, Stuttgart.

Newall, C. A., Anderson, L. A., and Phillipson, J. D. (1996). *Herbal Medicines: A Guide for Health-Care Professionals,* The Pharmaceutical Press, London.

Oelze, F., Brinkmann, H., and Wiesenauer, M. (1994). *Naturheilverfahren bei Herz-Kreislauferkrankungen,* Hippokrates Verlag, Stuttgart.

Rätsch, C. (1998). *Enzyklopädie der psychoaktiven Pflanzen: Botanik, Ethnopharmakologie und Anwendung,* AT Verlag, Aarau, Wissenschaftliche Verlagsgesellschaft, Stuttgart.

Robbers, J. E. and Tyler, V. E. (1999). *Tyler's Herbs of Choice: The Therapeutic Use of Phytomedicinals,* The Haworth Press, Inc., Binghamton, New York.

Saller, R., Reichling, J., and Hellenbrecht, D. (1995). *Phytotherapie,* Karl F. Haug Verlag, Heidelberg.

Schilcher, H. (1997). *Phytotherapy in Paediatrics,* Medpharm Scientific Publishers, Stuttgart.

Schilcher, H. (1999). *Phytotherapie in der Kinderheilkunde,* Third Edition, Wissenschaftliche Verlagsgesellschaft, Stuttgart.

Schilcher, H. and Kammerer, S. (2000). *Leitfaden Phytotherapie,* Urban und Fischer, Stuttgart.

Schneider, G. and Hiller, K. (1999). *Arzneidrogen,* Fourth Edition, Spektrum Akademischer Verlag, Heidelberg.

Schulz, V., Hänsel, R., and Tyler, V. E. (2001). *Rational Phytotherapy: A Physician's Guide to Herbal Medicine,* Fourth Edition, Springer Verlag, Berlin, New York.

Tang, W. and Eisenbrand, G. (1992). *Chinese Drugs of Plant Origin,* Springer Verlag, Berlin, New York.

Teuscher, E. (2003). *Gewürzdrogen,* Wissenschaftliche Verlagsgesellschaft, Stuttgart.

Teuscher, E. and Lindequist, U. (1994). *Biogene Gifte,* Second Edition, Gustav Fischer Verlag, Stuttgart.

Teuscher, E., Melzig, M., and Lindequist, U. (2004). *Biogene Arzneimittel,* Sixth Edition, Wissenschaftliche Verlagsgesellschaft, Stuttgart.

Tyler, L. (2000). *Understanding Alternative Medicine: New Health Paths in America,* The Haworth Press, Inc., Binghamton, New York.

Tyler, V. E. (1993). *The Honest Herbal: A Sensible Guide to the Use of Herbs and Related Remedies,* Third Edition, The Haworth Press, Inc., Binghamton, New York.

Tyler, V. E. (1994). *Herbs of Choice: The Therapeutic Use of Phytomedicinals,* The Haworth Press, Inc., Binghamton, New York.

Wagner, H. (1999). *Arzneidrogen und ihre Inhaltsstoffe, Pharmazeutische Biologie,* Volume 2, Sixth Edition, Wissenschaftliche Verlagsgesellschaft, Stuttgart.

FOREWORD

As a new millennium begins, **THE EVENTFUL CENTURY** series presents the vast panorama of the last hundred years—a century that has witnessed the transition from horse-drawn transport to space travel, and from the first telephones to the information superhighway.

THE EVENTFUL CENTURY chronicles epoch-making events like the outbreak of the two world wars, the Russian Revolution and the rise and fall of communism. But major events are only part of this glittering kaleidoscope. It also describes the everyday background—the way people lived, how they worked, what they ate and drank, how much they earned, the way they spent their leisure time, the books they read, and the crimes, scandals and unsolved mysteries that set them talking. Here are fads and crazes like the Hula-Hoop and Rubik's Cube . . . fashions like the New Look and the miniskirt . . . breakthroughs in entertainment, such as the birth of the movies . . . and marvels of modern architecture and engineering.

MILESTONES OF MEDICINE charts the dramatic breakthroughs—such as the discovery of penicillin, the first successful heart transplant, the birth of the first test-tube baby and the eradication of small-pox—and the host of less celebrated but no less significant advances in medical science that have transformed people's daily lives, and their very expectations of life, during the 20th century. From the fight against infectious diseases such as diphtheria and tuberculosis in the early part of the century to the development of antibiotics in the second half, from early X-rays to computerized scanning, from blood tests to genetic testing, from the early kidney dialysis machines to heart pacemakers, technology and medical science have combined to produce continuous advances in the diagnosis and treatment of disease and to alleviate or cure many previously untreatable conditions. At the same time vaccinations, replacement hip operations, keyhole surgery, contact lenses and the contraceptive pill have become routine features of people's lives.

THE
EVENTFUL CENTURY
20th

MILESTONES
OF MEDICINE

Reader's
Digest

The Reader's Digest Association, Inc.
Pleasantville, New York/Montreal

MILESTONES OF MEDICINE
Edited and designed by Toucan Books Limited
Written by David Burnie
Edited by Helen Douglas-Cooper
Designed by Bradbury and Williams
Picture research by Christine Vincent

FOR THE AMERICAN EDITION
Produced by The Reference Works, Inc.
Director Harold Rabinowitz
Editors Geoffrey Upton, Lorraine Martindale
Production Antler DesignWorks
Director Bob Antler

FOR READER'S DIGEST
Group Editorial Director Fred DuBose
Senior Editor Susan Randol
Senior Designers Carol Nehring, Judith Carmel
Production Technology Manager Douglas A. Croll
Art Production Coordinator Jennifer R. Tokarski

READER'S DIGEST ILLUSTRATED REFERENCE BOOKS
Editor-in-Chief Christopher Cavanaugh
Art Director Joan Mazzeo

First English edition copyright © 1998
The Reader's Digest Association Limited,
11 Westferry Circus, Canary Wharf,
London E14 4HE

Copyright © 2000
Reader's Digest Association, Inc.
Reader's Digest Road
Pleasantville, NY 10570

Copyright © 2000
The Reader's Digest Association (Canada) Ltd.
Copyright © 2000
The Reader's Digest Association Far East Limited
Philippine copyright © 2000
Reader's Digest Association Far East Limited

Library of Congress
Cataloging in Publication Data:
Milestones of medicine.
    p. cm.—(The eventful 20th century)
    ISBN 0-7621-0285-3
        1. Medicine—History—20th century.
            I. Reader's Digest Association. II. Series.

    R149 .M556 2000
    610'.9'04—dc21
                                            00-023050

FRONT COVER
From Top: An assortment of drugs available today;
Dr. Christiaan Barnard; Artificially colored X-rays of
hands.

BACK COVER
From Top: Newborn babies; Imaging and treatment
technology; A surgeon at work.

Page 3 (from left to right): Tilting disc valve for heart;
a foil pack of Prozac; Med-E-Jet inoculation gun; a
nurse pouring out medicine, c.1900.

Background pictures:
Page 13: Red blood cells in sickle cell anemia
Page 37: Assorted pharmaceutical pills
Page 95: Colored X-ray of an arthritic hand
Page 129: X-ray diffraction crystallography

Address any comments about Milestones of
Medicine to
Reader's Digest, Editor-in-Chief, U.S. Illustrated
Reference Books,
Reader's Digest Road, Pleasantville, NY 10570

To order additional copies of Milestones of
Medicine, call 1-800-846-2100

You can also visit us on the World Wide Web at:
www.readersdigest.com

CONTENTS

CHANGES IN MEDICINE

IN THE 20TH CENTURY, SCIENTIFIC PROGRESS AND TECHNOLOGICAL INNOVATION REVOLUTIONIZED THE ANCIENT ART OF HEALING

In 1913 the first domestic refrigerator went on sale in Chicago, and Henry Ford introduced the world's first industrial assembly line. The Mexican president and the King of Greece were both assassinated, and there were signs of the international tensions that would soon trigger the First World War.

In the same year, the first attempts at diagnosing breast cancer using X-rays were made, and the world's earliest artificial kidney had its trial run. But as far as human health was concerned, the launch of the refrigerator ranked among the year's greatest events. Although no one foresaw it at the time, it would become a powerful weapon in the fight to improve public health.

In these early years of the 20th century, medicine reached a decisive moment. For the first time in human history, doctors had a better than 50-50 chance of helping their patients get well, despite the fact that, by modern stan-

HEALTHY CHILL In 1900, insulated ice cupboards provided the only reliable way of keeping food from going bad in hot weather.

dards, the average doctor was poorly equipped. Most of the drugs and compounds then in use dated back many years, and most had only generalized effects. But doctors also had access to a limited number of new and highly specific remedies for potentially fatal

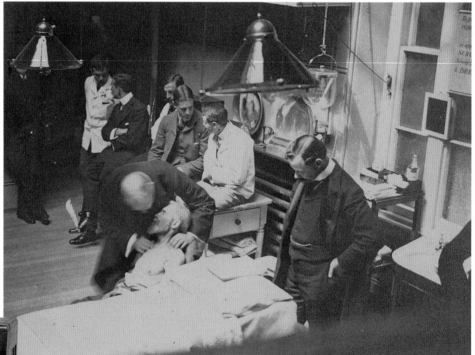

PROFESSIONAL SCRUTINY While their junior colleagues take a break, two doctors examine a patient at a London hospital in 1905.

diseases. One of these was an antitoxin that helped to combat diphtheria—then a major childhood killer. Another was a drug called Neosalvarsan, which was used as a treatment for syphilis, still a widespread disease in 1913. Unlike the other contents of a doctor's bag, these two remedies seemed almost magical in their results.

In the early 20th century, doctors may have looked much like their predecessors of a few decades before, even if they did not dress in quite the same way

and even though some drove cars. But scientific research had started to make a major impact on their profession.

Preparing the path
History rarely creates neat divisions between one century and the next. This is certainly true in medicine, because early 20th-century medicine owed a great deal to discoveries made in the years immediately before. The discovery of X-rays, for example, which was announced early in 1896, had a tremendous impact on medical diagnosis once the new century began. So did the discovery that mosquitoes spread malaria, which was made in 1897, and the discovery that rats spread bubonic plague, which was made in 1898. In a completely different field of medicine, another crucial event occurred in 1893 with the publication of *The Psychic Mechanism of Hysterical Phenomena*. Written by two Austrian physicians, Sigmund Freud and Joseph Breuer, it marked the beginning of psychoanalysis, a method of treating mental disorders that was to become supremely important in the new century. But of all the late 19th-century breakthroughs, the one that had by far the greatest influence on

medicine in the years that followed was the germ theory of disease.

A century ago, the connection between germs and disease was still a relatively recent discovery. The general public knew little about them, and some older doctors were not that much wiser. Germs, or bacteria, had first been observed in 1683 by a Dutch microscopist named Antonie van Leeuwenhoek. Leeuwenhoek's discoveries had no immediate effect on medicine because for a long time no one imagined that there might be any link between the microscopic world and the much larger world that humans inhabit. But in the mid 1860s all this changed. France's silk industry was in the grip of a disease that decimated silkworms, and the distinguished chemist Louis Pasteur was called in to find out what was wrong. Pasteur showed that the disease was caused by bacteria, and when he ordered the destruction of all infected silkworms and the food they had touched, the epidemic came to a halt. Pasteur went on to link bacteria with other diseases, not only those in animals, but also in human beings. He was joined in this work by Robert Koch, the German researcher who established modern medical bacteriology.

Not everyone was convinced by the idea that diseases were caused by living organisms.

INVISIBLE ENEMY Before the spread of disease was properly understood, children were particularly vulnerable, especially among the poor such as this immigrant family in the United States in 1910.

In 1892, for example, a German named Joseph von Pettenkofer swallowed some water containing cholera bacteria, and by a stroke of luck managed to avoid catching cholera himself. But this dramatic gesture flew in the face of an ever-growing body of evidence showing that the germ theory was correct. Germ theory explained, for example, why operations became safer when Joseph Lister, an English surgeon, began to use antiseptic sprays on his patients. It explained why improvements in water supplies and sewerage helped to cut the incidence of disease, and it also explained why—when it was eventually launched—the refrigerator, by inhibiting the growth of bacteria in food, proved so useful in preventing illness. By 1900, germ theory was paramount in medical thinking. To some researchers it seemed possible that behind every disease, including cancer, there lurked a microorganism waiting to be discovered.

The discoveries of Pasteur and Koch gave birth to the science of immunology—the study of resistance to disease—and a few decades later produced the search for antibiotic drugs. Both developments were examples of a new kind of medicine—one that saw disease as something that could be explored, analyzed and, with appropriate measures, even wiped out altogether.

Changing practices

In the 19th century, doctors thought of their profession as the "healing art." By the 1920s, that art was already turning into a science, but change affected different practitioners in different ways.

In the 19th century, many people looked to their family doctor as the final authority on all medical matters. However, even in the late 1890s, the pace of change meant that it was becoming increasingly difficult for the ordinary doctor to keep up. So much new knowledge was being generated that individual doctors could not possibly master it all, and as a result they increasingly referred patients to someone else.

This trend toward specialization began in North America and was well under way by the time the new century began. In cities, general practitioners were soon hard to find, and patients went straight to the specialist instead. In most European countries, specialization never went this far, and the traditional doctor remained the first port of call in the health-care system. When a second opinion was needed, a visit to a consultant was prescribed.

The public demand for specialities such as neurosurgery and psychiatry rocketed, although some other fields fared less well. For example, venereology—

MARKETING MEDICINE A pharmacy in London (right), and a drugstore in the U.S. In the early 1900s, the word "drug" had few of the illicit connotations it has today.

the branch of medicine that deals with sexually transmitted disease—saw a boom in business immediately after the First World War, but a slow period a decade later. This was due partly to the use of new drug treatments for syphilis, and partly to increasingly effective public health campaigns. In the 1920s, one long-established venereologist on London's Harley Street anticipated this trend, and minimized the effect on his income by subletting some of his rooms. The maneuver proved so profitable that many other doctors followed suit, creating a scramble for prestigious Harley Street addresses that has lasted ever since.

In rural practices, change was often much slower. On both sides of the Atlantic, many country doctors continued to mix and dispense medicines themselves, acting as a combination of doctor and pharmacist. Before the First World War, mass-produced tablets were still relatively uncommon, and the doctor or his assistant would make pills by hand, grinding up powders with a small amount of adhesive gum, and then putting the mixture into a special roller to turn out the finished pills. In the United States, where the rural population was often much more scattered than in Europe, doctors sometimes doubled as proprietors of drug stores, which sold not only drugs, but also eyeglasses, toiletries, surgical trusses and a

MEDICAL RECORDS

At the beginning of the 20th century, patients' medical records were remarkably brief. In 1900, for example, a man admitted to the Pennsylvania Hospital had his case summed up on a single page, despite the fact that he was in the hospital for three months. Since then, patients' records have expanded enormously. This change has come about partly through a proliferation of diagnostic data, and partly through an increase in the number of specialists involved in assessment and treatment.

wide range of household goods. In the days before routine abdominal surgery, the surgical truss—a kind of specially designed corset—was the main form of treatment for hernia, and was used widely.

Communication, particularly in rural practices, was not nearly as easy as it is today. In the early decades of the 20th century most patients did not have telephones, and even some doctors had reservations about

RADIOACTIVE REMEDIES

DANGEROUS WATERS William Bailey's radioactive water gave purchasers more than they had bargained for.

As generations of entrepreneurs have realized, fortunes can be made by creating popular medicines and patent remedies—regardless of whether or not they work. The 19th century saw a boom in this branch of the health business, but even in the more enlightened 20th century, patent remedies have had an enthusiastic following.

One of the most dangerous of these proprietary medicines went on sale in 1925. Known as Radithor, it was the brainchild of William J. Bailey, a shady American businessman who had already been convicted of manufacturing a fraudulent cure for impotence. Bailey decided to cash in on the public's fascination with radioactivity, which in the 1920s was a new and much misunderstood phenomenon. He had some success with "radioactive harnesses," which he claimed could cure all sorts of metabolic disorders, but with Radithor he struck gold. As its label proclaimed, Radithor was radioactive water that contained tiny amounts of radium. More questionably, advertisements put out by the Bailey Radium Company guaranteed that Radithor was "harmless in every respect." In an early example of the mass mailing, Bailey sent a promotional pamphlet to every registered doctor in the United States. Headed *Radithor, the New Weapon of Medical Science*, the pamphlet extolled the virtues of the new medicine, and promised large discounts on all orders.

As a business venture, the results were impressive. Radithor was soon selling at the rate of nearly 100,000 bottles a year. Bailey sent cases to friends and acquaintances, and even gave it to the racehorses he had acquired with his new wealth. When doubts were raised about the safety of radium, he pointed to the fact that he took Radithor himself, and that he was in robust good health. However, not everyone shared Bailey's confidence in his product. In 1930, the Food and Drug Administration warned of the dangers of radioactive substances, and in 1931, an investigator visited the home of Eben M. Byers, a millionaire businessman who had taken more than 1,000 bottles of Radithor in four years. Byers was severely ill, and the investigator's findings stopped Radithor sales in their tracks.

"A more gruesome experience in a more gorgeous setting would be hard to imagine," the investigator wrote. "We went to Southampton where Byers had a magnificent home. There we discovered him in a condition which beggars description. Young in years and mentally alert, he could hardly speak. His head was swathed in bandages. He had undergone two successive jaw operations and his whole upper jaw, excepting two front teeth, and most of his lower jaw had been removed. All the remaining bony tissue of his body was slowly disintegrating, and holes were actually forming in his skull." Not only were his bones and teeth radioactive, his breath was radioactive as well. He was suffering from massive radium poisoning. Radithor was banned in December 1931, although this move came too late to save Byers, who died just three months later. But the Radithor incident did alter the regulation of patent remedies. Until then, medicines were assumed safe until they proved themselves otherwise. Radithor showed how dangerous this assumption could be.

LETHAL LEGACY Today, decades after they were first used, even empty bottles of Radithor are still dangerously radioactive.

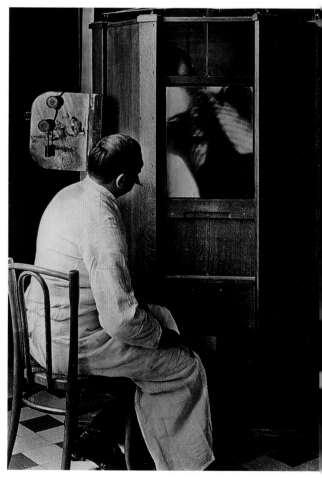

ADVERTISING ASPIRIN A Dutch car extols the virtues of aspirin in 1929. The manufacturer—Bayer—was already a household name.

register, which was controlled by the members of the profession. Medical registration did not necessarily ensure that a doctor was competent—particularly as some had received only the most basic training—but it did at least mean that there were recognized standards. Registration also helped to shape other medical professions such as dentistry, ophthalmology and pharmacology, which at one time were subject to few controls.

The professionals

With professionalization came increased status and rewards. By 1900, doctors had grown used to being restrained when it came to advertising their services, and even the most eminent practitioners had only a discreet nameplate on their consulting room doors. But in allied medical professions, business practice was not always so subtle. In the United States particularly, competition between dentists became so fierce that dental associations were forced to step in. In Connecticut, a directive was issued in 1930 to stamp out the large signs above dentists' offices. The directive stipulated that lettering could be no higher than 4 inches on the ground floor and 6 inches on the second.

Pharmacists were in a slightly different position. Although they faced restrictions in advertising themselves, there was a tradition of advertising pharmaceutical products. Despite gradually tightening controls, this tradition never died out. In Europe, the old-fashioned chemist shop, with its rows of jars and bottles bearing Latin inscriptions, was transformed by a flood of ready-made medicines. Out went the jars and bottles that contained the pharmacist's basic ingredients, and in their place came

SEE-THROUGH CABINET A French doctor examines a patient's chest on a fluorescing screen.

ready-made drugs. The pharmacist also changed from someone involved in making up prescriptions to someone who sold proprietary medicines and counted out tablets, and who kept a watchful eye on what each patient was prescribed.

Once registration was in force, it became much easier to identify and eliminate some of the quacks and charlatans who had dogged 19th-century medicine—although it also restricted more respectable alternative therapists. In the United States, practitioners of alternative therapies were often pursued through the courts, and several prominent figures eventually ended up in jail. In Britain, meanwhile, the Chief Medical Officer reacted scornfully to the idea that herbalists might be given some form of legal recognition, saying in 1923 that he doubted whether a trained herbalist was any less dangerous than an untrained one.

Sometimes new developments brought changes that could not be ignored. When X-rays came into use in medicine, it was by no means obvious whose job it would be to take them. X-rays used photographic plates, so

them, suspecting that they might compromise the doctor's duty of confidentiality. This meant that doctors had to be called to urgent cases by word of mouth, or by hastily scribbled messages describing a patient's symptoms. House calls were generally far more common than today, particularly during epidemics of diseases such as diphtheria, when each young patient might need visiting three times a day.

In the days before antibiotics, if a dangerous infectious disease was diagnosed, rapid action was needed to ensure that the disease did not spread. Usually this meant banning

KEEP TAKING THE TABLOIDS

Medicines in tablet form first appeared in the 1840s, but they did not supplant pills until the early 1900s. Unlike pills, tablets—or "tabloids," as they were also called—are made by compressing dry powders so hard that the grains stick together. Pills are made by mixing powdered ingredients with an adhesive paste.

visitors and keeping children away from school, but sometimes more drastic action was called for. In one practice in Sanford, North Carolina, smallpox patients were sent to a remote building in the woods surrounding the town. The patients were nursed by a man who had already had the disease, and a red flag warned other people to stay away.

By 1900, all doctors in North America and Europe had to be listed on a medical

A DOCTOR CALLS

CARRYING THEIR BLACK BAGS LIKE BADGES OF OFFICE, DOCTORS MAKING HOUSE CALLS WERE ONCE A FAMILIAR SIGHT IN ALMOST EVERY NEIGHBORHOOD

How would a doctor's home visit in 1910 compare with one 50 years later? The changes are many, and they start not at the bedside but at the front door. In 1910, many doctors still arrived on horseback, and the vast majority of them were men. Half a century later horse transport was a distant memory, and so too was the mainly male medical profession. Following the upheavals and social changes of the Second World War, many more women went to medical school, and the result in the 1960s was a steep rise in the number of women doctors.

As the doctor reaches the patient's bedside, several other differences are apparent. In 1910, diagnosis often involved detailed questioning of the patient, backed up by a limited number of physical tests. These included palpation (examination by touch), auscultation (use of a stethoscope to check the condition of the heart and

more comfortable, but did not tackle the underlying problem. In these early years of the century, the doctor's bag contained some powerful plant-based painkillers and sedatives, such as morphine and cocaine, but very little that could help infections. Instead, the doctor aimed at building up the patient's strength so patients could fight off the illness themselves. As a result, emphasis was placed on diet, and on not "overtaxing" the digestion. Fifty years later, palliation was no longer a medical priority. Instead, the accent was on cure, often with the help of antibiotics. Where rapid drug action was needed, a 1960s doctor would give injections—something that was rarely done in 1910.

In cases of accidents and emergencies, doctors in 1910 were not nearly as well-equipped as their counterparts 50 years later, but they were more used to relying on their own resources. Most doctors in 1910 had extensive experience in setting fractured bones and dealing with wounds, in their offices or in patients' homes. By 1960, improved communications and faster ambulances meant that most of this kind of treatment was carried out in hospitals.

RAPID RESPONSE In 1900, urgent medical help arrived at the speed of a trotting horse. Today, air ambulances allow medical teams to be on site within minutes.

lungs), and taking the patient's temperature. By 1960, these diagnostic tests had been joined by routine blood pressure measurement, which was still rare in 1910. Questioning the patient now took second place to a detailed examination.

In 1910, many patients would receive an accurate diagnosis of their problem, but the doctor would often be unable to give them anything more than reassurance and palliative treatment. This made the patient

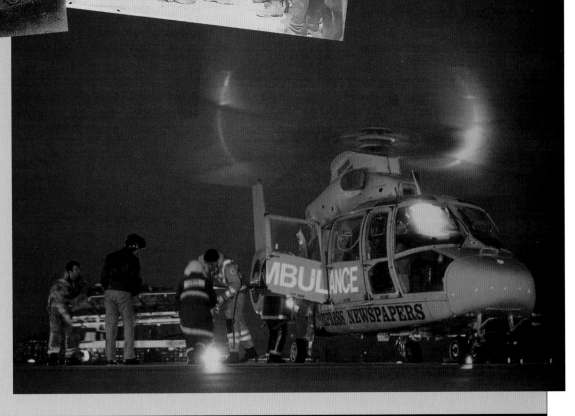

they needed the skills of a photographer. At the same time, a knowledge of anatomy was required by the person taking X-rays, and this was the province of the medical establishment. Eventually, a new medical profession—radiology—was born, although not before some lengthy arguments concerning exactly whose job it would be to interpret the pictures once they had been taken.

The business of good health

If a doctor from 1900 could be brought forward to the present day, the advances in treatment and technology would seem astounding. So, too, would the influence that

THINGS TO COME Housed in balloon-like shields, X-ray therapy units of 1919 (above) heralded the immensely more sophisticated imaging and treatment technology of today (right).

medicine now has on daily life.

One feature of this influence is the apparently unstoppable rise of medicine as an economic force. Not only has the number of doctors increased, but so has the number of people involved in organizing and carrying out medical care. They range from chemists and engineers working on the production of drugs and high-tech medical equipment to paramedics, nurses, physiotherapists, dieticians, social workers and computer systems analysts; not to mention health-care managers, whose responsibility it is to run services that often rival multinational businesses in size.

The medical profession and its allied industries now account for a large proportion of the economic activity in the developed world. They stimulate research and generate an unquenchable demand for new technology, new drugs and new ways of dealing with once intractable problems.

This alliance between medicine and industry dates back to the 19th century and the pioneering work of organic chemists in Germany. Since then, it has expanded into something that touches the lives of almost everybody on our planet, with the exception only of those in the poorest countries. Its story appears to be one of continuing success in the fight against ill health, but as the expansion continues, some people have begun to question the direction that modern medicine is following. One danger is that the alliance between medicine and industry could become self-propelling, so that it serves its own

needs more than those of the public.

According to a report written by a former Director-General of the World Health Organization in 1977, this is exactly what has happened. "Most of the world's medical schools prepare doctors not to take care of the health of the people, but instead for a medical practice that is blind to anything but disease and the technology for dealing with it," the Director-General's report said. "The medical empire and its closely related aggressive industry of diagnostic and therapeutic weapons sometimes appears more of a threat than a contribution to health."

It is an outspoken analysis, and one that most recipients of replacement hips, heart pacemakers, antibiotics and genetically engineered insulin would find difficult to agree with. In the 20th century, medicine produced its share of misjudgments and changes of direction, but it also saw some of the most resounding successes in the human race's struggle for better health.

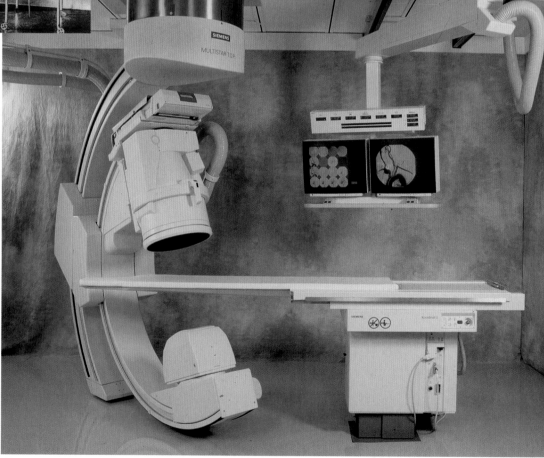

DISEASE AND DIAGNOSIS

DURING THE 20TH CENTURY, IMPROVEMENTS IN DIAGNOSTIC TESTS AND PROCEDURES REVOLUTIONIZED THE AGE-OLD STRUGGLE TO IDENTIFY AND OVERCOME DISEASE. WITH THE HELP OF NEW IMAGING TECHNIQUES, LABORATORY ANALYSIS AND WAYS OF MONITORING THE BODY'S ACTIVITY, DOCTORS TODAY HAVE SOMETHING THAT THEIR 19TH-CENTURY PREDECESSORS LACKED— A BATTERY OF RELIABLE AND OBJECTIVE TESTS ON WHICH EFFECTIVE TREATMENT CAN BE BASED.

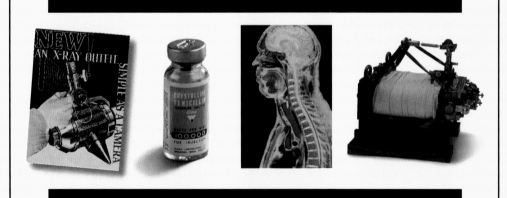

AIDS TO DIAGNOSIS

BLOOD PRESSURE CHECKS AND EKG RECORDINGS MARKED THE START OF A NEW PRECISION IN THE ANCIENT ART OF MEDICAL DIAGNOSTICS

In his best-selling book, *The Poet of the Breakfast Table*, published in 1872, the American writer and physician Oliver Wendell Holmes gave his thoughts on the value of medical diagnosis. As far as most patients were concerned, he wrote, falling ill and being examined was rather like being bitten by a dog and being given just one piece of information—the dog's name. Naming the culprit might satisfy the doctor, but often left the patient no better off.

His views echoed a commonly held sentiment of the day. Despite advances in diagnostic techniques, including the invention of the stethoscope and the clinical thermometer, doctors at the time were still powerless to combat most diseases and disorders, and there was something to be said for remaining in ignorance of the disease from which one was suffering. However, medicine was making progress in the 1870s, particularly in the fields of surgery and infectious disease, and by the time the 20th century arrived, accurate diagnosis was no longer simply something that satisfied doctors. In some cases, it could be the first step toward a successful cure.

As the 20th century opened, there was still some lingering debate about whether diagnosis should be considered an art or a branch of modern science. This debate stemmed from a time when doctors diagnosed diseases primarily by asking questions, rather than by giving patients detailed physical examinations. Diagnostic checks were few, and only rarely did a doctor touch a patient's unclothed body.

The invention of the stethoscope in 1816 by the French physician René Laënnec helped to break down the taboo of physical contact, but with female patients in particular, the taboo's power took some time to wane. Scurrilous cartoons hinted that doctors might use the stethoscope as an excuse for improper intimacy, and many women refused to be examined in this way. Even severe illness failed to make some patients relent. Queen Victoria, for example, was never given a full physical examination by the last of her doctors, despite the fact that she eventually became so sick that she died while under his care.

Royalty apart, however, attitudes were changing, and by 1900 physical examination formed an integral part of most consultations. The instruments involved were few by modern standards, and diagnosis still depended as much on the doctor's intuition and experience as on objective observation. But as the new century progressed, diagnostic techniques became much more precise. And specialized instruments and tests improved the accuracy of the doctor's analysis and increased the likelihood that treatment would have the required effect.

Measuring blood pressure

One of the most important aims of any physical examination is to assess a person's general state of health. A clinical thermometer can do this to a limited extent, but with the invention of the sphygmomanometer, doctors acquired a more powerful addition to the diagnostic armory.

The sphygmomanometer measures blood pressure. It evolved from an instrument made by the German physiologist Carl Ludwig in the mid-1850s that tapped into the bloodstream via a cannula, or narrow tube, inserted into an artery. Ludwig used an ingenious recording device to trace the rise and fall of blood pressure on a soot-covered drum, but the need to puncture an artery meant that his apparatus had no safe place in the doctor's repertoire. Later, another German, Karl Vierordt, made a conceptual breakthrough. He realized that there was no need to make actual contact with the bloodstream itself. First, an artery

LISTENING IN Working in a factory clinic, an American doctor in the 1920s uses a stethoscope to examine a patient.

1906 Einthoven records electrical activity of the heart

1910 Blood tests become routine

1926 First portable EKG machine

1930s Electron microscope First electronic EKG machine developed

had to be gently squeezed until it collapsed and its blood flow stopped. Next, the pressure was slowly released. There would then come a moment, as the blood started to push past the obstruction, when the pressure outside the artery was exactly equal to the pressure inside. By listening for this precise

GAUGING HEALTH   The modern sphygmomanometer (right) has changed little from early versions, such as this 1909 French one.

moment, the blood's pressure could be measured from outside the body. This became the principle behind the sphygmomanometer, which could measure blood pressure without causing a patient any harm.

The first sphygmomanometers—and indeed many still in use today—measured

NIKOLAI KOROTKOFF'S SOUNDS

In 1905, a Russian military surgeon, Nikolai Korotkoff, was the first person to analyze the sounds that can be heard through a stethoscope during a blood-pressure test. Now known as Korotkoff sounds, they are created by turbulent blood as air in a sphygmomanometer cuff is slowly released. Knowledge of these sounds allowed more accurate blood-pressure measurements to be made.

blood pressure by using a column of mercury in a glass tube. This column was originally connected to a water-filled pipe that ended in a flexible bulb, or "pelotte." The pelotte would be used to make an artery collapse, usually in the arm, and when the blood flow was allowed to resume, the height of the mercury column would give

the systolic pressure, the pressure in the arteries when the heart contracts, or beats.

Today, blood-pressure readings are still expressed in millimeters of mercury, even though many modern gauges use electronic displays or spring-activated dials instead of a mercury column. However, the water-filled pelotte is now a thing of the past. In 1896, the Italian physician Scipione Riva-Rocci devised a much more reliable way of interrupting blood flow, using a rubber sleeve that was wrapped around the whole arm and then filled with air. Riva-Rocci's sleeve was only about 2 inches wide, and in 1901 it was superseded by a more successful design, the now-familiar inflatable rubber cuff. By listening for the sounds produced by moving blood, doctors today take two measurements. The first is the systolic or peak pressure, and the second is the diastolic pressure, the blood's background pressure between each heartbeat.

A study based on more than 10,000 blood-pressure readings was published in 1902 by medical researchers in France. In 1915, an even larger study, based on more than 18,000 readings, was produced by the Prudential Life Insurance Company of Newark, New Jersey. Analysis of these results enabled the insurance industry to set "normal" values for blood pressure. Anyone who fell above these limits was deemed to be at greater risk of ill health, and therefore subject to increased premiums. This definition of normality remained unchanged until

1952, when a study of nearly 75,000 readings showed that the original limits were slightly too low. Today, the generally accepted norm is a maximum diastolic, or relaxed, pressure of 85 mmHg (millimeters of mercury). Anyone having a diastolic pressure above 95 mmHg is usually given treatment to return it to a safer level.

Monitoring the heart

Although Carl Ludwig's puncture method of measuring blood pressure had no lasting role in medical diagnostics, his other achievement—the recording in chart form of bodily changes—has had a tremendous impact in many branches of medicine, most notably in monitoring the heart. The soot-covered drum has given way to the computerized print-

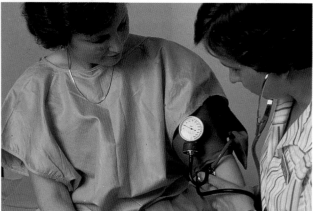

out and screen display, but the underlying principle—visual information that can be taken in at a glance—is still the same.

During the 20th century heart disease became increasingly prevalent, due both to increased longevity and, in affluent countries, to diets high in animal fats. Methods of diagnosing heart malfunction have therefore gained enormous significance. Since the early 1900s, several different techniques have been explored for charting the activity of the heart. One of the most unusual, called ballistocardiography, was actively investigated in the 1920s and 1930s. It worked by recording the rebound that occurs throughout the body after each heartbeat—a phenomenon that can sometimes be seen by standing absolutely still on sensitive bathroom scales. During a ballistocardiographic test, the person being examined lay on a specially constructed table that was free to sway horizontally. A set

1960s Stick test invented for urine analysis

WIRED An electrocardiograph in 1902 (below) used water baths to complete an electrical circuit that ran through the body. Today's compact EKG machines (right) use small electrodes fastened onto the skin.

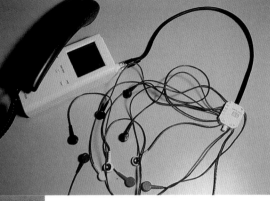

WIRED An electrocardiograph in 1902 (below) used water baths to complete an electrical circuit that ran through the body. Today's compact EKG machines (right) use small electrodes fastened onto the skin.

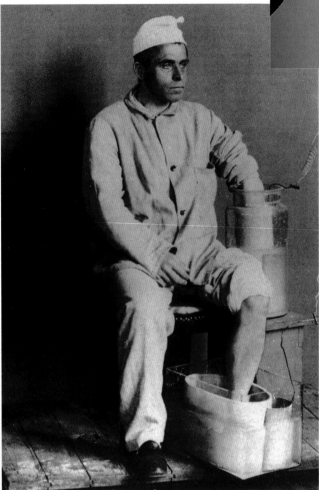

1930s saw the first electronic EKG, which used valves to amplify the incoming signals. By the late 1940s, EKGs had become small and straightforward enough to earn their place in the consulting room, and by the 1950s half of American general practitioners had one. Today—after the introduction of modern semiconductors—an EKG is so small that it can be strapped around a person's waist and used to compile a complete record of the heart's electrical activity around the clock.

Chemical analysis

Blood pressure and EKG checks are non-invasive ways of assessing a person's state of health, so-called because they do not involve any intrusion into the body itself. Other non-invasive diagnostic techniques include examining the body's waste products. This kind of examination has an ancient history. In medieval Europe, for example, doctors often claimed to be able to diagnose any disease purely by looking at a patient's urine—a technique called uroscopy.

Although hugely overblown, uroscopy was based on physical changes that do have some diagnostic value. For example, the presence of blood gives urine a reddish color, while infections can turn it cloudy. By the beginning of the 20th century, the arcane art of uroscopy had been reborn as the much more precise science of urology. Doctors in 1900 still used urine's color as a guide to disease, but they also employed several other tests, including the measurement of specific gravity. An unusually high specific gravity shows that a patient's urine contains more dissolved substances than it should, although on its own this test says nothing about what these

accompany each heartbeat. It has proved so effective that since the 1950s it has become an essential diagnostic tool.

The EKG was pioneered by the Dutch physiologist Willem Einthoven, who in 1901 managed to record one of the world's earliest graphs of heart activity. Einthoven's first machines were built just before the invention of electrical amplification, and a large part of his achievement lay in the way he managed to pick up the faintest electrical signals from the body and use them to produce a permanent recording of the beating heart. His first electrocardiogram (the word that describes the recording itself, rather than the machine) showed the steady and rhythmic profile of a typical heartbeat, with three electrical waves that follow each other in quick succession. In the decades after Einthoven's initial pioneering work, specialists and general practitioners discovered that they could deduce a huge amount of information about the health of a patient's heart by measuring the exact shape of the waves.

Einthoven's earliest machine weighed over 600 pounds, could not be moved, and required four or five people to operate it. However, progress was rapid. The first portable machine appeared in 1926, and the late

of detectors sensed the slight movement after each heartbeat, and built up a chart that showed the amount of blood expelled by the heart. This, doctors hoped, would reveal any signs of disease. A second method, called phonocardiography, worked by analyzing heart sounds. It proved useful for detecting heart murmurs and other abnormalities, but despite a considerable amount of research into it, the method declined in popularity in the 1960s.

The ballistocardiograph and phonocardiograph never gained a permanent place in the doctor's consulting room because they faced competition from a quite different kind of instrument, the electrocardiograph, or EKG. This machine works not by detecting movement or sound, but by picking up and amplifying the tiny electrical changes that

INSTANT TEST A urometer set made by Parke, Davis & Company in 1900. It carried out a simple chemical test to detect urea, a waste product in urine.

CHARTING THE HEART

TRAFFIC VIBRATION AND OVERHEATING WERE JUST TWO PROBLEMS THAT HAD TO BE OVERCOME IN CREATING A PERMANENT RECORD OF THE BEATING HEART

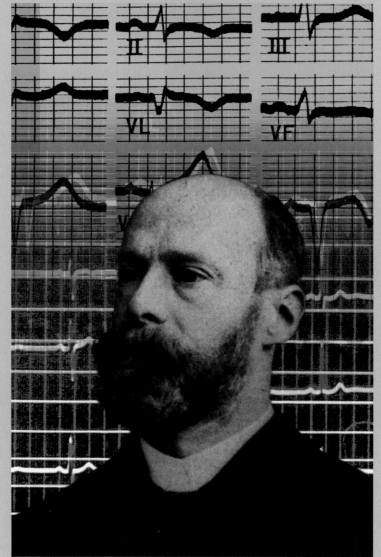

When the heart beats, it undergoes electrical changes that are almost imperceptible. For Willem Einthoven, who pioneered the electrocardiogram, or EKG, in the Dutch town of Leiden at the turn of the century, the measurement of such minuscule changes became a major challenge.

At first, Einthoven tried to tackle the problem by using a device called a capillary electrometer. This consists of a slender tube filled with mercury. The apparatus is arranged so that the mercury expands or contracts when electricity flows through it. The movement of the mercury is magnified by a set of lenses, and then projected onto a sheet of photographic paper. Despite its sensitivity, Einthoven had great difficulties with this delicate piece of apparatus. The mercury took too much time to react, and it was so susceptible to vibrations that it would be thrown wildly off course by horse-drawn traffic passing on the cobbled streets outside. In an effort to overcome this difficulty, Einthoven excavated the floor of his laboratory and had it filled with a layer of rock over 10 feet deep. The electrometer wobbled on.

Einthoven eventually succeeded by abandoning the capillary electrometer altogether in favor of a different kind of measuring device—a string galvanometer. The string in question was an extraordinarily fine strand of quartz, measuring just 0.0001 inch wide, made by attaching a quartz crystal to a bench at one end and to an arrow at the other. The crystal was then heated and, as the quartz melted, the arrow suddenly flew through the air, pulling a strand of quartz behind it. The strand was then plated with silver and used to suspend a tiny mirror between the poles of a large electromagnet. When the galvanometer picked up currents through the body, the string twisted slightly, turning the mirror. This deflected a beam of light, causing it to trace a shifting path on a moving photographic plate.

Unlike the capillary electrometer, this system worked. However, Einthoven's early machine was so gigantic that it filled two rooms and required constant cooling. There was no question of moving it out of his laboratory to carry out diagnostic trials, but one of Einthoven's former teachers came up with an ingenious alternative. By using a telephone line connecting his laboratory to a local hospital to transmit the electrical impulses from a patient's heart, Einthoven was able to carry out "telecardiograms"—the world's first form of diagnostic testing conducted over a distance. Through these electrocardiograms, Einthoven became familiar with many of the abnormalities that indicate heart disease, and his achievement was officially recognized in 1924 with the award of a Nobel prize.

During its almost century-long history, the EKG has become progressively simpler to operate, and has made use of a variety of display and recording media, from the photographic plate to the computer print-out. Probably the most famous and evocative of them all is the moving trace on an oscilloscope screen, which first flickered into being in 1952. Since then, the image of this moving trace has come to epitomize the role of technology in modern health care.

TRACES ON A PAGE The string galvanometer invented by Willem Einthoven (above) recorded the tiny electrical changes that occur each time the heart beats.

QUICK DIP In less than a minute, a multiple test stick performs seven separate chemical analyses on a specimen of urine.

substances are, or what their presence means.

Thanks to the work of scientists such as the German biochemist Otto Folin, early 20th-century doctors could probe much further than this. Folin studied tests used in German breweries, and in a neat translation of chemical expertise, he applied brewery techniques to the world of medicine. In 1905, Folin described new methods for analyzing urine for four important constituents—urea, ammonia, creatine and uric acid—unusual levels of which can indicate liver diseases, infections and kidney disorders. By the 1920s, checks like these could be carried out by the general practitioner in the office, giving a greater insight into what might be wrong.

One substance in particular—glucose—has great medical significance when present in urine. Glucose is the body's fuel, and in ordinary circumstances is never disposed of as waste. But in people suffering from diabetes, too much glucose often circulates in the blood, and the kidneys try to dispose of the surplus. In 1900, the underlying cause of diabetes was still unknown, but glucose in the urine was known to be a warning sign. In patients suffering severe diabetic

symptoms, a doctor—or sometimes the patient—would take daily urine glucose tests using a substance called Benedict's solution, which turns from bright blue to a rusty red when it is mixed with anything containing sugar. The patient's diet would then be carefully controlled until the sugar level dropped.

After the discovery of insulin in 1921, diabetes became much easier to control, but glucose measurement was still important. Benedict's solution gave way to tablet-based tests that were simpler to use at home, but in the 1960s a completely new kind of testing system became available—the test on a stick. A specially treated stick is immersed in a urine sample for a specific length of time, and the presence of certain substances, such as glucose or proteins, in the urine causes the stick to change color. Stick tests are now used for a wide variety of checks for metabolic disorders.

Testing blood

Compared to urine tests, routine blood tests are a much more recent development in medical diagnosis. The practice of removing blood for testing did not become widespread until the 20th century, partly because of a lack of safe instruments, and partly because doctors were becoming aware of the possibilities of cross-infection. In the early years of the century, tests revolved around the physical structure of the patient's blood, rather than its chemical make-up.

Although blood looks like a simple liquid, it actually consists of red and white cells suspended in a straw-colored fluid

called plasma. Red cells, which carry oxygen, are almost identical to each other, while white cells, which help to defend the body against infection, are much more diverse. There are millions of cells in a single drop of blood, with the red cells vastly outnumbering the white. However, as medical researchers soon discovered, the exact proportions fluctuate according to the body's state of health. Measurement of the numbers of cells can

THE PRESIDENT'S BLOOD

One of the earliest examples of a diagnostic blood test took place after President William McKinley was shot by an anarchist on September 6, 1901. A single bullet penetrated his stomach, creating grave concern about the possibility of blood poisoning. Six days after the attack, a blood test showed no signs of a rise in the white cell count, which suggested that the president might be on his way to recovery. However, the hope proved unjustified. An infection set in shortly afterward, and on the ninth day the president died.

therefore provide some important clues about a range of disorders.

Initially at least, it was a laborious task to make these measurements. One of the pioneers of hematology—the scientific study of blood—was the physiologist Karl Vierordt, who also worked on blood-pressure measurement. He took a week to estimate the red-cell count in a single sample of blood,

ANXIOUS MOMENT A German worker being given a blood test at the Junkers aircraft works in the early 1940s. Disposable syringes were not available so sterility was essential.

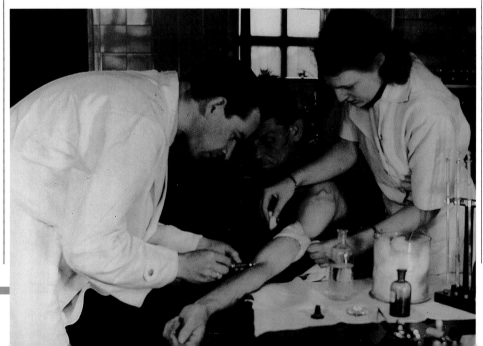

BLOOD CELLS The blood sample above shows a much greater volume of red cells than white (colored yellow here). A centrifuge, left, separates the blood cells from the plasma.

and experience. Because they saw cases of many infectious diseases on a regular basis, they were often so familiar with them that they felt no need for additional diagnostic help. By the outbreak of the First World War this independent approach was on the wane. By now, diagnostic bacteriology was a rapidly growing science. With suitable tests, infections could be identified at an earlier stage, which—even in the days before antibiotics—bought precious time in the fight against disease. However, most of these tests were no longer within the capabilities of the average general practitioner.

Before disease-causing bacteria could be identified, they usually had to be cultured, which involved growing them on a suitable medium in exactly the right conditions. For example, in a case of suspected tetanus, the blood sample had to be kept away from oxygen, because the bacterium responsible for tetanus, called *Clostridium tetani*, could

which would not have impressed busy doctors. However, Vierordt was working by eye and counting cells on a microscope slide. By the time blood tests became common in general practice, from about 1910 on, a much faster method had been developed. A sample of blood was placed in a tube and spun around in a centrifuge so that the cells and plasma separated. When the tube was removed from the centrifuge, the red cell volume could then be read off on a scale. This method, which is still used today, takes just a few minutes and gives a figure called a hematocrit. A low hematocrit is often a sign of anemia, while a high figure can reveal disorders of the liver or kidneys.

Hematocrits—or blood counts— are sometimes carried out in the doctor's office, but today, as with many clinical tests, they are more often conducted in laboratories that specialize in medical analysis. This shift away from the doctor's premises has been a consistent feature of 20th-century diagnosis, and nowhere has it been more marked than in the field of infectious disease.

Diagnosis in the lab

At the beginning of the century, very few doctors took blood tests to identify infectious diseases. Some had the skills and equipment necessary to identify disease-causing bacteria under a microscope, but this kind of diagnostic technique was strictly for the enthusiast. Instead, most doctors relied on outward symptoms

THE DISCOVERY OF SICKLE CELL ANEMIA

The early years of blood tests shed light on several previously unknown or misunderstood medical conditions. One of these was sickle cell anemia. In the mid-1900s, while examining a sick black student from the West Indies, James Herrick of Chicago decided to carry out a blood test on his patient. Looking down a microscope, he discovered to his surprise that many of the man's red blood cells had lost their normal coin-like shape, and had become slender and curved like sickles. Herrick reported his findings to the Association of American Physicians in May 1910. Over the next few years, it became apparent that his patient was not the only one suffering from this disorder. Instead, it turned out to be widespread in the black population. It was also discovered in people of Mediterranean origin, but was otherwise very rare. The disease is, in fact, hereditary, which is why it is concentrated in specific population groups.

In the years since Herrick's discovery, physiologists have tracked down the underlying cause of the disorder. It is triggered by a genetic abnormality in hemoglobin, the chemical pigment that red cells use to carry oxygen. In the finest blood vessels, where oxygen is relatively scarce, molecules of the abnormal hemoglobin crystallize, giving the red blood cells their sickle shape. These cells often break apart, clogging blood vessels and causing damage to internal organs.

SICKLING CELLS The distorted, sickle-shaped red blood cells first observed by James Herrick can break up and obstruct small blood vessels.

MEDICINE AND MICROSCOPES

ELECTRONIC EYE A scientist uses a transmission electron microscope to study influenza viruses. This kind of microscope works by firing electrons through an ultra-thin slice of tissue.

At the beginning of the 20th century, the microscope was an exclusively optical instrument. High-quality lenses were available, but medical microscopists still faced the problem that many of the structures they wanted to observe—such as white blood cells and bacteria—were almost transparent, making details hard to pick out.

The main technique used to get around this difficulty was based on selective stains. These substances, which were developed from the middle of the 19th century on, are absorbed to different degrees by different types of cells. New stains created during the 20th century have helped in diagnostic work, particularly in the identification of cancerous tissue. Some stains are used specifically to reveal bacteria. The Gram stain, for example, devised by the Danish bacteriologist Hans Christian Gram in 1884, splits bacteria into two overall groups, helping in their identification. In the 1940s, bacteriologists discovered that these two groups—called Gram-negative and Gram-positive bacteria—react differently to antibiotics, making the stain even more useful in diagnosis.

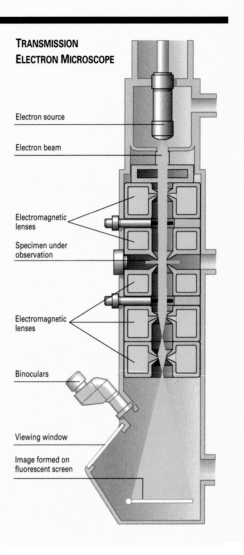

TRANSMISSION ELECTRON MICROSCOPE

Electron source

Electron beam

Electromagnetic lenses

Specimen under observation

Electromagnetic lenses

Binoculars

Viewing window

Image formed on fluorescent screen

The most powerful light microscopes magnify up to 2,000 times, which is powerful enough to reveal details of most kinds of cells. However, until the invention of the electron microscope in the 1930s, the smallest agents of disease—viruses—could be detected but not seen. With the development of the electron microscope after the Second World War, the first images of viruses became available. The huge magnifying power of these new microscopes—some modern types are 100 times more powerful than a light microscope—also enabled researchers to investigate the detailed structures within individual cells.

Electron microscopes now play an important part in medical research, but light microscopes and selective stains remain the pathologist's most useful tools for everyday diagnosis.

grow only in oxygen-free conditions. This is no easy matter for a doctor to arrange in the office. Neither is the lengthy business of preparing and staining microscope slides so that bacteria can be observed. As a result, doctors withdrew from this kind of work, and the modern pathological laboratory—or "path lab"—came about.

Before the 1930s, most laboratory tests were carried out manually, limiting the number that could be processed. As automation became more widespread, whole batches of tests could be carried out simultaneously. In addition, improved techniques meant that tests became more sensitive. Each sample could be divided up many times and tested in many different ways. Tests for infectious diseases no longer relied on spotting elusive

BUG HUNT A bacteriologist at work in a laboratory in the 1930s. The glass-topped box reduces the risk of contamination.

microorganisms. Instead, immunological methods were developed that could identify chemical clues circulating in the blood—chemicals produced either by the microorganisms, or by the body in its attempts to resist attack. These tests give results in minutes and can often be done with one pinprick of blood.

With this kind of technical support, today's doctors work in a very different way than those of 80 or 90 years ago. Then, when doctors examined their patients, diagnosis was often a matter of fitting each person into an illness or disorder that the doctor already knew. Now, with the help of laboratory tests, doctors can identify conditions that they may never have encountered before, or even be aware of. Intuition no longer plays as great a part as it once did, generally to the benefit of the patient. As one American surgeon remarked in the early years of the century, "Diagnosis by intuition is a rapid method of reaching a wrong conclusion."

IMAGING THE BODY

THE DISCOVERY OF X-RAYS CREATED A NEW BRANCH OF MEDICINE AND HERALDED A CENTURY OF BREAKTHROUGHS IN IMAGING TECHNIQUES

On January 5, 1896, a newspaper in the Austrian capital, Vienna, carried news of a sensational discovery. It reported that a German physicist, Wilhelm Roentgen, had stumbled upon a form of radiation capable of traveling through the human body and photographing the bones within. Unlike other breakthroughs of the time, this one generated an instant and almost insatiable curiosity among scientists and the general public. Startling pictures of hands taken with Roentgen rays—or X-rays as Roentgen preferred to call them—

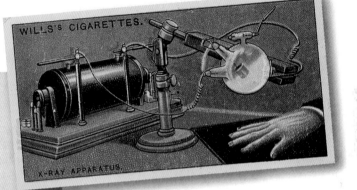

appeared in all kinds of publications, from medical journals to billboard advertisements, and medical equipment companies vied with each other to produce their own versions of Roentgen's original apparatus. Roentgen himself became an instant celebrity. He was showered with awards that included the first Nobel prize for Physics, and also the Prussian Order of the Crown, presented personally by Kaiser Wilhelm II.

The diagnostic value of X-rays quickly became apparent. Within weeks, Viennese doctors had

SEEING WITHIN An X-ray from 1905 (left) reveals a toy bicycle lodged in a child's throat. By 1915 even cigarette cards trumpeted the new X-ray technology (above).

taken the first medical radiographs, showing a hand injured by gunshot and a badly set fracture of the forearm. Several months later, X-ray evidence made its first appearance in a court of law. The court, sitting in Denver, heard how a leading surgeon had failed to diagnose a boy's broken thigh after the young patient had fallen off a ladder. The jury was presented with X-ray evidence of the boy's injury, taken by a photographer who had a keen interest in "roentgenology."

Although Roentgen's chance discovery fell a few years short of the 20th century, it proved to be one of the greatest contributions to medical advances in the new century. Until X-rays appeared as a diagnostic tool, doctors had

SIMPLICITY ITSELF This 1930s advertising brochure announces an improved dental X-ray machine—simple enough for any practitioner.

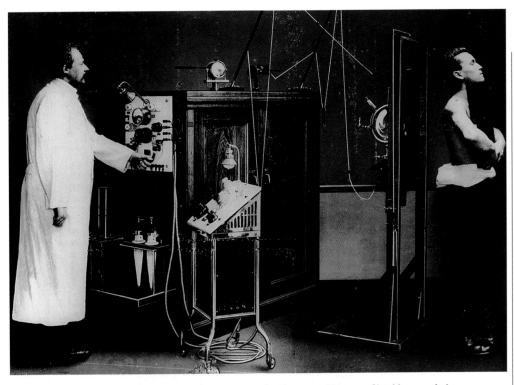

REVEALING MOMENT Clutching a sealed photographic film, a patient stands still while an X-ray is taken of his chest.

it were being squeezed. By applying a voltage that fluctuated very rapidly—some thousands of times a second—Langevin made crystals change shape so quickly that they generated sound. The sound was too high to be audible, but this was exactly what Langevin wanted. Audible sounds have relatively low frequencies and they spread out across a wide range, whereas high-frequency sounds spread over a narrower range. They can be aimed like a searchlight beam, and if they hit anything, they are reflected with a minimum of scattering.

By linking a crystal transmitter to a detector, Langevin successfully used this kind of sound to locate objects submerged in a laboratory tank. His first transmitter consumed a colossal amount of power—enough to kill nearby fish—but the results were encouraging. For the first time, sound had been used to reveal what the eye could not see. While Langevin was carrying out his work, the tide of war turned and an armistice was declared. But although his detection system came too

almost no means of examining the interior of the human body without first cutting it open. The further development of X-rays through improvements in image quality produced by cathode tubes and image intensifiers, and the discovery of some of their dangers

When a voltage is applied to a crystal, the crystal changes shape, creating ultrasound.

A LUCKY OVERSIGHT LEADS TO GALLBLADDER X-RAYS

In the early 1920s, an American medical intern named Warren H. Cole was trying to produce the first X-ray images of the gallbladder by giving dogs doses of iodine. The work was time-consuming and repetitive, and seemed to be going nowhere. Suddenly, after hundreds of failures, one picture produced a perfect result:

"I called Dr. Graham, who was working late as usual. We stood there admiring the dripping film with a white blob in the center, as if we had found a treasure chest of gold."

But a bewildered Dr. Cole had no idea why this single X-ray had worked.

"Finally, in desperation, I called on Bill [the animal caretaker] and asked him if the treatment of the dog had been in any way different to that of the other three dogs. When I told him that this dog showed exactly the thing we were looking for, his expression of apprehension changed, and [he] meekly stated: 'Well, Dr . Cole, there *was* one thing different. I forgot to feed that dog the morning he was injected.'"

Cole immediately realized that lack of food was the key, because the gallbladder empties during a meal. Thanks to a lucky oversight by the animal caretaker, gallbladder X-rays soon became routine.

and shortcomings, stimulated research into other ways of seeing into the body. Some flourished but then faded away; others are at the forefront of medicine today.

Seeing with sound

Unlike the discovery of X-rays, the development of ultrasound scanning had its origins in warfare, rather than in the pursuit of abstract physics. During the First World War, German submarines posed a threat to shipping in the North Atlantic, and the French government asked the physicist Paul Langevin to investigate ways of detecting submarines from the surface.

At this time, it was known that crystals developed an electrical voltage if pressure was applied to them. (This phenomenon, known as the piezoelectric effect, later became widely used as a way of lighting gas cookers.) It was also known that the effect worked in reverse: if the voltage was applied to the crystal, the crystal changed shape, as if

Ultrasound, emitted by a crystal transmitter, is reflected by underwater objects and by fetuses.

SOUND SCAN Paul Langevin's work led to imaging by sound.

DISCOVERING THE DANGERS OF X-RAYS

X-RAY IMAGING TOOK THE MEDICAL WORLD BY STORM, BUT EVIDENCE SLOWLY CAME TO LIGHT OF ITS POTENTIALLY LETHAL EFFECTS

In the wave of excitement that greeted the discovery of X-rays, little thought was given to the possibility that they might have harmful effects. Initially, far more attention was given to other aspects, such as the imagined threat that X-rays posed to personal privacy. On both sides of the Atlantic, enterprising businesses advertised such products as "X-ray proof underwear," which would protect ladies from X-ray–equipped peeping toms. Others offered to supply X-ray pictures taken through walls and doors as evidence in divorce cases.

The misunderstandings that surrounded X-rays were not confined to the general public. Initially, no steps were taken to shield radiographers from the rays. Worse still, many used their hands to test the "hardness" of the rays when setting up an X-ray tube and fluorescent screen. The harder or stronger the rays, the more intense the shadows they threw.

This habit soon produced the first symptoms of a new illness, which became known as X-ray hand. At first, the disease showed itself as little more than a blister, but unlike a true blister this one did not heal. In most cases it continued to grow, and often became a cancerous growth that spread across the hand and wrist. The effects could be devastating. In 1908, one of its victims, John Hall-Edwards, presented a paper on the disease to Britain's Royal Society of Medicine. He showed a series of photographs and X-rays of his own hands and wrists as they were progressively eaten up by cancer, before and after his hands were finally amputated. Patients who had been exposed to large X-ray doses also experienced a range of unwelcome effects, from hair loss to badly burned skin. Some soon died, while others succumbed to X-ray induced cancer years or even decades later.

The experience of radiographers and patients alike clearly demanded action. Early on, Roentgen had discovered the blocking effect of dense metals such as lead, and these became a first line of defense. Many early radiographers took to wearing metal-lined gloves and aprons, and some also used metal helmets. X-ray rooms, viewed through lead glass windows, were suggested as another means of reducing exposure.

As technology improved, methods were found to screen X-ray tubes so that their radiation could be more accurately directed. At the same time, government stepped in. The concept of the maximum permissible dose was introduced in the United States in the 1940s, and as the years passed, permissible levels were steadily stiffened around the world. In the late 1950s, foot X-ray machines—a novelty seen in some shoe stores—disappeared after a study linked X-rays to childhood cancer. This event finally spelled the end of X-rays outside the scientific and medical world.

EARLY PRECAUTIONS   Lead-lined gloves and aprons offered some protection to pioneer radiologists.

**PROFESSIONAL HAZARD
A French doctor's hands show the effects of radiation burns.**

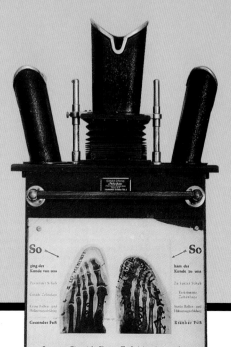

STEP THIS WAY   The notice on this 1930s German "Pedoscope" invites customers shopping for shoes to have an X-ray free of charge.

So →
ging der
Kunde von uns

← So
kam der
Kunde zu uns

Lassen Sie sich Ihren Fuß hier kostenlos
durchleuchten

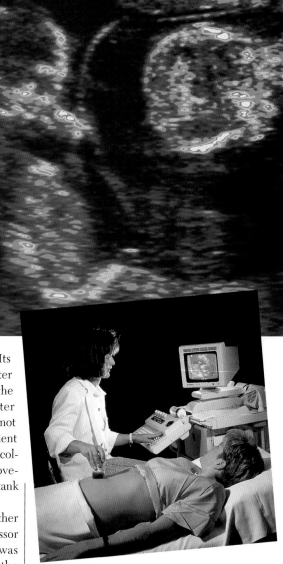

late to play any part in the First World War, it was crucially important in the Second World War, and also in the very different field of medicine.

Using transmitters much smaller than Langevin's formidable prototype, physicists discovered that ultrasound was reflected not only by the surface of solid objects, but by any kind of boundary inside them. These boundaries included invisible cracks within pieces of machinery, and also the boundaries between different types of living tissue. By the 1930s, Austrian neurologists succeeded

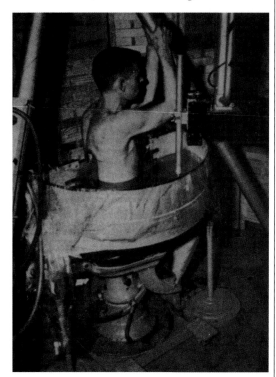

TOTAL IMMERSION A volunteer sits wedged into Douglass Howry's "somascope"—an early attempt to use sound to see through the body.

in passing a beam of ultrasound through the skull and produced rudimentary images of what lay inside. In 1949, scientists in Argentina used ultrasound to reveal kidney stones, and in 1952, Douglass Howry, an American radiologist, published an ultrasound "slice" through an arm, showing the skin, bones and some hints of the muscles.

Later in the 1950s, Howry and his colleagues developed a device called a "somascope," which could scan the chest and neck. The somascope was as much a triumph of make-do mechanics as it was of medical physics. It consisted of a modified dentist's

chair topped by a semi-circular water-filled bath, which was made from part of the gun turret of a B-29 bomber. The patient sat in the chair, his or her trunk wedged firmly into a slot in the bath, while the ultrasound scanner traveled around a rail inside the bath below the water line. The result, although far from perfect, was one of the earliest imaging systems to reveal a part of the body from many different angles.

Despite its ingenuity, Howry's somascope proved something of a disappointment. Its chief drawback was that it relied on water to "couple" the ultrasound waves into the body, as any air between the transmitter and the skin acts as a reflector. But it is not a simple matter to surround part of a patient with a pool of water. As Howry and his colleagues discovered, one misjudged movement could result in the contents of a tank gushing over the laboratory floor.

On the other side of the Atlantic, another pioneer of ultrasound, Ian Donald, Professor of Midwifery at Glasgow University, was contemplating similar problems. From the early days of radiography, it was clear that images of the developing fetus could be valuable in assessing growth and diagnosing disorders. However, X-rays could trigger dangerous side effects, such as childhood leukemia. Ultrasound seemed a promising way out of this dilemma and Donald started to use it on his patients. Early scans were produced using rubber-bottomed tanks, which sat directly on top of the mother's body, or—if things went wrong—slipped off. But after a series of experiments, a simpler technique emerged: instead of using water, the ultrasound transmitter could be "coupled" to the patient by using a layer of oil. Smeared over the skin, this allowed a hand-held probe to scan deep within the womb.

Today, thanks to this technique, ultrasound routinely produces the most arresting medical images of all: the first sight of a new life, cocooned within the safety of its mother's body. Unlike X-rays, the procedure has

ECHOES FROM WITHIN Ultrasound is one of the safest forms of imaging. The scan at the top shows twin fetuses within the womb.

no dangerous side effects for the mother or her child.

The internal eye

Unlike X-rays or sound waves, which pass through soft body tissue, light is strongly absorbed by body tissue and can pass through no more than a few millimeters of skin. The only way to see into the body's interior is to shine light either through one of its natural openings, or through a surgical incision, and the instrument used to do this is the endoscope.

The earliest endoscopes included devices such as the retinoscope and otoscope, which are used to examine the eyes and ears. Until the late 1800s, their light source was a naked flame, usually a candle or spirit lamp, mak-

1909 Endoscope
used to view a
patient's stomach

1949 Ultrasound
used for
diagnosis

ing them cumbersome to use. By 1900, naked flames had been replaced by miniature electric bulbs. These were small enough to be fixed to the tip of an endoscope, rather than the base, so that the light source could be pushed inside the body. The bulbs were often hot enough to burn living tissue, and

TOILET-TISSUE TRACINGS

Some of the most important observations in radiography were made on an unlikely medium—toilet tissue, placed over a fluorescent X-ray screen while X-raying patients. The "artist" was Walter Cannon, an American physiologist. In the early 1900s, Cannon added barium sulphate to food to make the human stomach and intestines show up when looked at under X-rays, and his tracings showed how food is moved along by peristalsis—a process of muscular contraction. Among Cannon's discoveries was that emotions affect the rate at which food moves through the body. He published his studies in 1911.

so some designs included a cooling system.

In 1909, a surgeon succeeded in using a light-bearing endoscope to view a patient's stomach, but the process was a difficult one. The instrument was completely rigid, and had to be guided down the patient's throat and esophagus entirely by touch. Despite

INNER VISION A surgeon guides a laparoscope (a type of endoscope) into a patient's body through a small incision in the abdomen.

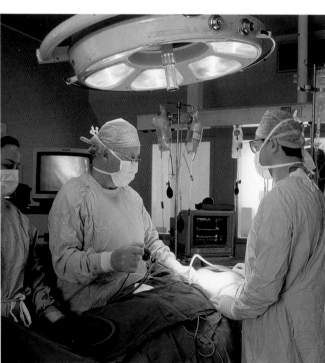

some ingenious advances involving flexible tubing and mirrors in the 1920s and 1930s, endoscopes needed skill and patience to use.

A technical advance then offered a new way forward. In the 1950s, physicists started to investigate the possibility of using slender fibers of glass to carry beams of light. A phenomenon called total internal reflection ensures that light shining in at one end shines out of the other, whether the fiber is straight or bent. Although these fibers are made of glass, they are so slender that they can be bent without breaking.

Fiber optics transformed the endoscope, allowing much deeper probing into the body. By the mid-1960s, the cable-controlled flexible endoscope had appeared. One bundle of fibers is used to carry light to the endoscope's tip, while another bundle carries light back to an eyepiece, so that the scene can be viewed. By operating handheld controls, a surgeon can guide an endoscope through the most complex internal spaces, from the fluid-filled capsules around joints to the interior of blood vessels.

Feeling the heat

In the 1960s, just as ultrasound became a practical possibility, another form of non-intrusive imaging emerged onto the medical stage: thermography, which works by recording the temperature pattern on the surface of the body. In a healthy person, the core temperature—the body's temperature in regions well beneath the skin—is a constant 98.6° Fahrenheit. By contrast, the surface temperature varies from place to place. The chest, for example, is almost as warm as the core, but the fingertips—particularly in cold weather—can be several degrees colder. Superimposed on these overall patterns are small but significant differences caused by localized sources of heat. Inflammation, for example, produces a local rise in temperature, as does rapid cell division which occurs

THERMAL IMAGE In this computer-processed thermogram of a man, the warmest areas are white and the coolest dark blue or black.

1950   2000

1950s Light-carrying endoscope produced using fiber optics

1958 Ultrasound first used to diagnose fetal disorders

1960s Thermography used for body imaging

1972 Computerized axial tomography (CAT) introduced commercially for medical imaging

1980 Positron emission tomography (PET) Magnetic resonance imaging (MRI)

when a piece of tissue becomes cancerous. This means that, in theory at least, thermography can show how the body is working.

During the 1930s and 1940s, researchers on both sides of the Atlantic investigated skin temperature as a possible clue to disorders and diseases. It was a cumbersome business. The measurements were made by sticking dozens of heat sensors onto the patient's skin, a process that could take several hours to complete. But in the early 1950s, military technology provided another way of making these measurements. Instead of measuring temperature directly, it could now be sensed by infrared detectors—the kind of instruments that were being developed to enable soldiers to see in the dark.

Infrared detectors work by collecting the invisible infrared light, or heat rays, given off by all warm objects. Detectors produced in the late 1950s could sense temperature differences no smaller than about 4° Fahrenheit at best—easily enough to show enemy soldiers lurking in the darkness, but not sufficiently discriminating to reveal the small temperature differences found over the human body. But the technology was advancing quickly and its medical supporters were enthusiastic.

In an atmosphere of excitement and expectation, it seemed that thermography might well replace X-rays in the early detection of breast cancer and other major diseases. "Thermography will add a significant new dimension to the diagnosis and prognosis of disease," announced a contributor to *Scientific American* magazine in 1967. But a series of lengthy clinical trials in the late 1960s and early 1970s were disappointing. Within overall limits, the body's surface temperature is just too variable. A trained radiographer can quickly spot a fracture revealed by an X-ray, but the information in a thermograph is much harder to interpret. Each person has their own pattern of temperature variations, which makes it very difficult to use hot or cold spots as warnings of possible disease. Researchers concluded that, on its own, thermography was not reliable enough as a diagnostic method.

Images of the brain

Given their potential for harmful side effects, new imaging methods have to be used with caution. Pioneers of X-rays tested their "Roentgenoscopes" on all kinds of non-living biological materials, from human corpses to plucked chickens, but when they subjected patients to the same high doses of radiation, the results were sometimes lethal.

By the time the 20th century reached its halfway mark, the dangers of X-rays were well-known, but their potential had yet to be fully exploited. This was because X-ray pictures, or radiographs, were still essentially shadow pictures: anything casting a heavy shadow, such as dense bone, masked whatev-

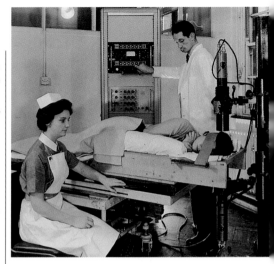

CANCER SCAN In the early 1960s, a patient with a suspected brain tumor has been injected with a radioactive isotope, which will pinpoint the tumor on an X-ray.

er lay behind it, while things that cast faint shadows were practically invisible. An X-ray picture of a plucked chicken exhibited in Berlin in 1896 showed this problem clearly: its principal bones were readily visible, but few of its internal organs could be seen.

X-ray pictures suffer from another problem. Seen singly, they convey no hint of depth, so a doctor has to use prior knowledge and experience to judge whether something lies near the surface of the body or much deeper within it. When that something is a bullet or a malignant growth, the judgment becomes supremely important.

This problem was recognized in the in-

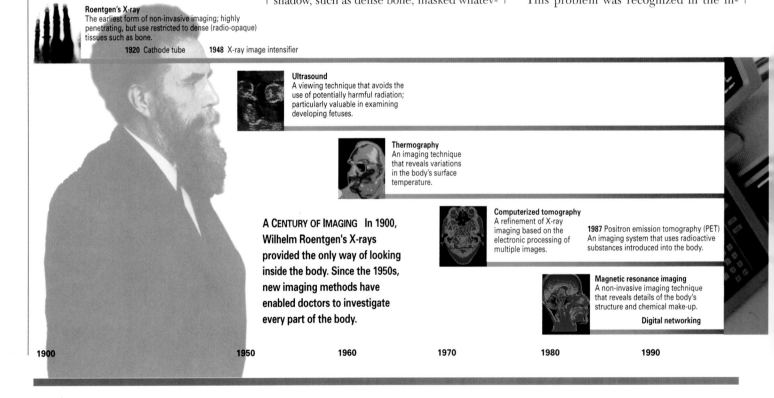

Roentgen's X-ray
The earliest form of non-invasive imaging; highly penetrating, but use restricted to dense (radio-opaque) tissues such as bone.

1920 Cathode tube   **1948** X-ray image intensifier

Ultrasound
A viewing technique that avoids the use of potentially harmful radiation; particularly valuable in examining developing fetuses.

Thermography
An imaging technique that reveals variations in the body's surface temperature.

A CENTURY OF IMAGING In 1900, Wilhelm Roentgen's X-rays provided the only way of looking inside the body. Since the 1950s, new imaging methods have enabled doctors to investigate every part of the body.

Computerized tomography
A refinement of X-ray imaging based on the electronic processing of multiple images.

1987 Positron emission tomography (PET) An imaging system that uses radioactive substances introduced into the body.

Magnetic resonance imaging
A non-invasive imaging technique that reveals details of the body's structure and chemical make-up.

Digital networking

1900   1950   1960   1970   1980   1990

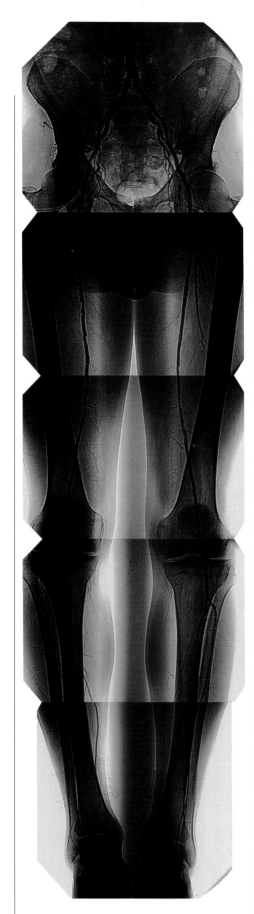

X-RAYED ARTERIES This composite angiograph shows all the major arteries in a patient's legs.

fancy of radiology, and by the 1920s the first attempts were under way to refine the use of X-rays so that they could look into the body rather than through it. This new kind of imaging was called tomography, from the Greek word meaning a slice. It worked by using an X-ray source and film that moved in parallel but opposite directions, with the patient's body sandwiched between them. This arrangement meant that any part of the body exactly halfway between the X-ray source and film stayed in focus throughout, while those at other depths were blurred.

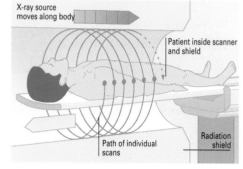

BODY SCAN A CAT scanner that spirals around the body as the patient moves along can build up an image of the bones and internal organs.

These early tomographs were of limited use, mainly because considerable expertise was required to distinguish the focused part of the picture from the unfocused details mixed with it, and in the 1920s and 1930s the only instrument that could interpret this jumble of information was the human eye.

Computerized axial tomography (CAT)

In the early 1960s, a leading California neurologist, W.H. Oldendorf, published a research paper highlighting the current difficulties in imaging the brain. Simple X-ray pictures disclosed few details of this mysterious organ, and the only available methods of probing further carried significant risks.

In the most common technique, called angiography, a dye containing a substance opaque to X-rays was injected into the bloodstream, usually at the neck. As the blood swirled up into the brain, the dye would travel with it, throwing the brain's blood vessels into sharp relief. A second technique, ventriculography, involved injecting air into the brain's ventricles—the interconnected spaces normally filled with a special fluid. By investigating the shape of the ventricles, it was sometimes possible to spot abnormalities in the surrounding brain.

Angiography was—and still is—useful for showing damage in blood vessels, and is often used to investigate the state of arteries around the heart. But because the dye remains in the blood vessels, it reveals little of the brain tissue itself. What was really needed, Oldendorf reasoned, was a non-invasive way of imaging the brain—a method of building up X-ray cross-sections that would allow the entire brain, including its softest

SLICES OF LIFE Godfrey Hounsfield's work with tomography made possible images such as this cross-section through the skull. The nose and eyes are at the top.

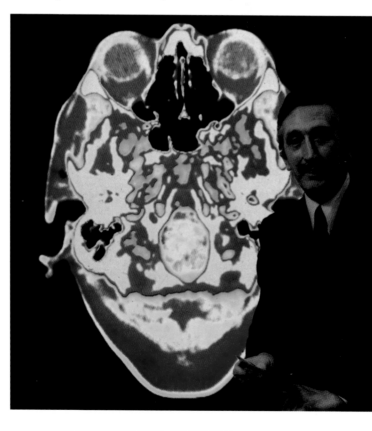

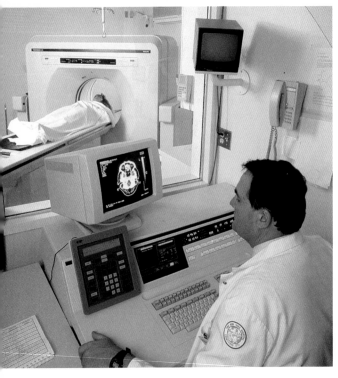

FULL DISCLOSURE An operator studies an image of a patient's brain during a scan. CAT scanners reveal tumors and blood clots that would once have gone undetected.

tissues, to be explored and mapped.

It was a highly ambitious proposal. Photographic films, the radiographer's traditional stock-in-trade, had to be replaced by sensitive X-ray detectors producing electronic signals that could be processed by computer. The X-ray source and detector had to be arranged so that they stepped their way in a line across the entire width of the patient's head, before turning through a small angle and starting again.

Oldendorf built a prototype of this kind of scanning system, but was unable to obtain the financial backing to take it any further. Instead Godfrey Hounsfield, a British electrical engineer, eventually managed to put this new form of imaging into practice. Hounsfield realised that a major problem with computerized tomography was the enormous amount of information that would be generated by the scanning process. Somehow a computer had to take in all this information, and then process it so that a complete image could be created.

Using techniques that have since become an everyday part of computing, Hounsfield found ways of eliminating redundant data. Instead of storing all the signals and then processing them once the scan was complete, the comput-

er worked by making an initial "guess" at the result, and modifying it each time a batch of signals came in. It was rather like someone predicting the whole of a company's annual accounts, and then updating them as the monthly results arrived. While this ingenious programming trick simplified the problem, the task of producing a result was still prodigious for the relatively primitive electronics of the time. In one of the first trials in the late 1960s, a test scan took nine days to complete. Hounsfield pressed on, however, and by the time details of the new scanner were announced in 1972, a complete cross-section of a patient's head could be produced in just five minutes, with the computer handling information from nearly 30,000 separate readings. The machine was received with a frenzy of enthusiasm.

By the mid-1970s, computerized axial tomography, known as CAT scanning, had moved on to the point where whole-body imaging was on the horizon, and hospitals were scrambling to acquire CAT machines. Despite the cost—a 1970s CAT scanner was nearly $400,000—supply initially struggled to keep pace with demand, as the number of scanners in the United States alone rose from 45 in 1974 to more than 900 in 1977.

Magnetic resonance imaging

Before computers arrived in medicine, hospital radiography departments had a unique character. Hidden away in row upon row of brown envelopes, processed films formed mute histories of accidents, injuries and false alarms, of setbacks and recoveries.

Despite the skill needed to interpret this kind of evidence, X-ray films provided visible images. But with the invention of computer-processed imaging in the 1970s, a profound change took place. Instead of being stored on film, medical images became something much less tangible—a sequence of numbers stored in a computer memory. Now they could be reprocessed at the touch of a button. This meant that images were no longer snapshots seen from the same angle as that from which they were taken. Instead, they could be used to assemble completely new views of the inside of the body.

It was into this very different world that the most recent imaging technique was born. Originally known as nuclear magnetic reso-

THE THREE-DIMENSIONAL BODY

In December 1994, an executed American murderer named Joseph Jernigan made posthumous history when his fully imaged body became accessible to researchers and medical students via the Internet. Jernigan donated his body to science, and after his death his body was scanned by X-rays and MRI, then sliced up into 1,878 sections, each 0.04 inch thick. Each of the body sections was then photographed, and the imaging information processed by computer to allow the entire body to be visualized from any angle. The complete set of data produced by the different scans totaled more than 42 gigabytes—about the same as storing approximately 10 billion words.

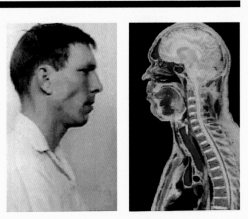

ELECTRONIC EPITAPH The body of Joseph Jernigan was converted into a digital image; one section (above right) shows the brain and spinal cord. A slice through his body (left) shows the upper torso and arms.

PET SCANNING

To make a CAT scan, radiation is passed through the body and then detected to form an image. But in one of the most recent forms of scanning, called PET or positron emission tomography, the source of radiation is actually introduced into the body.

PET scanning was developed in the late 1980s. During this type of scan, the subject is given a tiny amount of a substance, such as glucose, that has been chemically tagged or "labeled" with radioactive atoms. These atoms give off subatomic particles, called positrons, which collide with nearby atoms to produce gamma rays. A ring of sensors around the body detects these rays, enabling the labeled substance to be located very precisely.

The advantage of PET scanning over most other methods of imaging is that it shows the body's chemistry at work. For example, the brain uses large amounts of glucose to function. If radioactively labeled glucose is injected into the blood, the exact amount being used by different parts of the brain can then be seen. This reveals abnormalities such as areas affected by strokes, because in these, glucose uptake is much less than it would normally be.

The radioactive substances that are used in PET scanning are not dangerous because they have very short half-lives (the time taken for the radioactivity in a substance to be reduced by half through radioactive decay). Within a few hours, the radioactive substances have decayed into harmless atoms. This process of radioactive decay cannot be slowed down, however, so the source of the radioactive substances—a machine called a cyclotron—must be positioned close to the place where the scanning is carried out.

nance, or NMR, it has nothing to do with nuclear radiation and was renamed magnetic resonance imaging, or MRI, which is both more accurate and less alarming.

MRI had its roots in experiments carried out in 1946 when two American physicists, Felix Bloch and Edward Purcell, independently observed nuclear magnetic resonance in the laboratory. To bring about this process, a test material is put into a strong magnetic field and bombarded with microwave radiation. Some of the atomic nuclei in the material act as tiny magnets, lining up with the magnetic field. When the field is switched off, they return to their original positions, emitting a burst of radio waves as they do so. If these radio waves are used to create an image, they reveal both the shape of the object and something about its chemical make-up.

Attempts to use this form of resonance for medical imaging were first carried out in the early 1970s. The first MRI scanner to be tested on a living specimen had a working space just over 1 inch across—just big enough to fit around a mouse. Researchers soon realized that if a scanner was to deal with a human body, the magnets needed would be enormous. In addition, the mag-

netic field had to be even. Steel girders or iron pipes in the room housing the scanner would ruin the result.

For several years, different teams worked to overcome these problems. In the United States, one group, headed by Raymond Damadian, decided to build a giant electro-

magnet that worked by superconduction, whereby electricity flows almost unimpeded through very cold materials. The scanner, which the group named "Indomitable," took weeks to construct, and when it was complete one of Damadian's students slid inside it. For nearly five hours his body moved back and forth while a scan of his chest was completed. The result was a success, and by the time a scan of a human brain was published a year later, the new method of probing into the human body was exciting as much attention as had CAT not long before.

Unlike X-ray pictures of sharply outlined bones captured from a single angle at one moment in time, MRI scans brim with information. Because they do not depend on differences in density, they are capable of revealing the complete fabric of the body, from the enamel on teeth, which is almost as hard as steel, to the fluid-filled spaces around many internal organs. Today's radiologists are only just starting to grasp the usefulness of all this information.

NEW TECHNOLOGY Felix Bloch (left) and Edward Purcell (right) shared a Nobel prize in 1952 for their work on nuclear magnetic resonance, which gave rise to a new method of medical imaging.

EMERGENCY ACTION

IMPROVEMENTS IN FIRST AID, FROM BETTER BANDAGES TO BLOOD TRANSFUSIONS, HAVE SAVED HUGE NUMBERS OF LIVES

For the ordinary person in the year 1900, life held as many dangers as it does today. A glance through the medical periodicals of the time reveals a catalog of accidental injuries and the methods of emergency treatment. There were new suggestions for ways to clean wounds, improvements in techniques for dealing with fractures, and debate about the best methods for reviving people who had apparently drowned. One method, still advocated in the early 1900s, involved rolling the victim face down over a barrel so that air was forced into and out of the lungs.

In an age when there was still relatively little legislation about safety at work, industrial accidents were more common than now. Among the "laboring classes" on both sides of the Atlantic, who were most exposed to this kind of risk, disfigurement was rife because surgeons had yet to devise the procedures necessary to regraft severed parts of the body. As well as these risks, there were newer threats. City physicians recorded an increasing incidence of accidental death by electrocution now that houses were being connected to the power lines, and there was growing concern about traffic accidents, which for the first time were involving automobiles rather than horse-drawn carriages. Added to this list of hazards was one of the oldest of all: the effects of war.

Bandages and broken bones

Throughout the 20th century, war proved a grim testing ground for the effectiveness of first aid. During the First World War the motorized field ambulance made its appearance, but ordinary soldiers were often the first to tend to their wounded comrades.

As the century began, bandages and dressings were often difficult to use, partly because they were rarely of a standard shape or size. One of the first people to tackle this problem was a German military surgeon, Friedrich von Esmarch, who died in 1908. Esmarch devised the triangular bandage, which is still used today. The bandage formed

FIRST AID Early crepe bandages came with pictorial instructions.

EMERGENCY, 1925 An accident victim is strapped into a stretcher before being loaded onto a London ambulance.

1900 | 1950

1901 First manual heart massage

1905 First successful blood transfusion

1914 First blood transfusion using treated blood

1920s First sticking bandages

1930s Intravenous saline infusions given to accident victims

1935 First blood banks set up

1940s Antihistamine drugs introduced

1947 Electric shock used to restart a patient's heart

part of a standard first-aid kit, which was enclosed in rubberized cloth and was small enough to be stowed in a knapsack.

During the First World War, most bandages and dressings were made of cotton, but an unusual absorbent material found its way into use. This was sphagnum moss, a plant that grows in peat bogs and waterlogged moorlands throughout northern Europe. Pound for pound, dried sphagnum can absorb three times as much as cotton—and because it is acidic, it can also slow down the growth of bacteria. Although demand dwindled after the war, sphagnum dressings still remained in use in civilian life, and were being made as late as 1941.

The first sticking bandages appeared in the 19th century, when an American firm began to experiment with adhesives based on natural rubber. However, these bandages did not become widely available until the 1920s, and before then, dressings were usually fastened in place with knots or pins. In cases of serious injury, it was a problem to remove a dressing once its work was done as those made

First Aid in War British, French and Italian troops on the Western Front in 1918 (left). Soldiers carried field dressings (above) in their kit to tend wounded comrades during battle.

of absorbent materials, such as cotton, often became stuck to healing wounds. When a dressing was eventually removed, some of the newly formed tissue was likely to come with it.

The answer to this problem lay in a dressing that would absorb fluid but not stick to a wound. From 1914 on, various attempts were made toward this elusive goal. The first person to achieve real success was Louis Lumière, a French surgeon, who tried coating cotton gauze with a mixture of plant resin and paraffin wax. The result, called *tulle gras*, or waxed gauze, did not stick, and because it had an open weave, it still allowed the backing material to absorb fluid from a wound. Lumière's breakthrough took time to become recognized, but it paved the way for a range of dressings with non-stick surfaces. These stayed in use up until the 1960s, when impregnated gauze was superseded by synthetic non-stick liners of the type used today.

For accidents involving broken bones as well as cuts and bruises, careful diagnosis is essential before treatment can proceed. Nowadays, X-rays are used to pinpoint fractures, but until the years immediately after the First World War, when X-rays became routine, older diagnostic techniques often had to be employed. A doctor or medical officer would locate the fracture by careful manipulation of the injury, and if necessary would "reduce" the break by shifting the broken bones back to their correct positions. Once this procedure had been completed, the limb would be immobilized in a splint or in plaster of Paris until the broken ends knitted together.

Back from the dead

In cases of severe injury, whether on the battlefield or in civilian life, first aid has to be swift and methodical. Nowhere is this more true than in the gravest situation of all—when the victim's breathing and heartbeat both appear to have stopped.

Various forms of artificial ventilation, or artificial respiration, such as pressing against the victim's chest or diaphragm, or mouth-to-mouth treatment, have been practiced since Biblical times, but in 1900 there was still little agreement about which method worked best.

During the early 1800s, some forward-thinking doctors owned "resuscitation boxes," which included pipes and bellows that could be used to inflate a patient's lungs. A century later, in 1907, the German engineer Bernard Draeger made the first machine designed to carry out this work
continued on page 34

1955 External defibrillator used successfully for the first time

1960s Portable defibrillators introduced

1970s American helicopters used as ambulances during the Vietnam War

MEDICINE AT WAR

DURING THE 20TH CENTURY, MEDICINE HAD TO COPE WITH THE HORRIFIC EFFECTS OF MODERN WARFARE AND ITS IMPACT ON SOLDIERS AT THE FRONT

Throughout the 20th century, changes in medical knowledge and organization had a profound effect on soldiers and civilians swept up in war.

As the century opened, the British army was embroiled in a three-year conflict with the South African Boers. Following their experimental use in the Sudan, portable X-ray machines were brought into the campaign, but the new technology posed almost as many problems as it solved. At the time, X-ray pictures were taken on glass plates instead of films, and because the plates broke easily, the apparatus had to be kept well behind the front lines. After their long journey back from the front, injured soldiers often arrived for X-rays with their wounds already healed, and with bullets still in place.

By the time the First World War began in 1914, a far greater range of medical tests and treatments were available behind the lines. They included mobile laboratories for carrying out bacteriological tests, and surgical stations with improved X-ray machines situated much closer to the front. In the Allied armies, intravenous saline was used to treat shock, but blood transfusions took place only on a very limited scale.

Despite improved understanding about the causes of infection, outdated treatment techniques initially added to the already

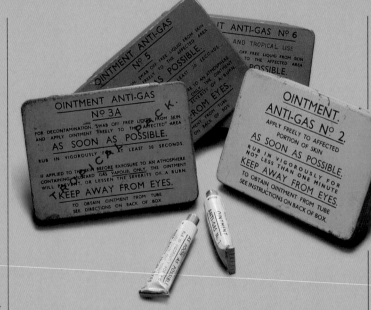

MEDICAL SUPPLIES By the Second World War, a decontamination ointment had been developed for use by British troops in a nerve-gas attack.

led to septicemia (blood poisoning) or gangrene, which could rapidly prove fatal. When the appropriate lessons were learned, and débridement was carried out more thoroughly, fatalities from infections dropped sharply, despite the lack of antibiotics.

The next major conflict in Europe, the Spanish Civil War, saw two new methods of controlling infection: the introduction of sulpha drugs (synthetic drugs containing sulphur) and the use of a new tetanus vaccine, which had been developed in France. This was also the first full-scale war in which stored blood was available for transfusions at the front.

By the time the Second World War broke out, medical organization, as well as treatment, had been transformed. Medical teams were now fully motorized, which meant that they were able to keep up with troops fighting at the front. Rapid movement of casualties also helped to improve survival rates. In Allied forces in Europe, a fifth of all casualties requiring surgery were operated on within 6 hours of being injured, and the bulk of the remainder within 12 hours. Wounds were routinely tested for bacterial contamination, and were cleaned and irrigated with sterile saline to prevent infection. During 1942, the first large-scale transfusions took place on the Allied side using stored blood. German medical teams had not yet perfected techniques for storing blood, so in the German forces most transfusions were still direct.

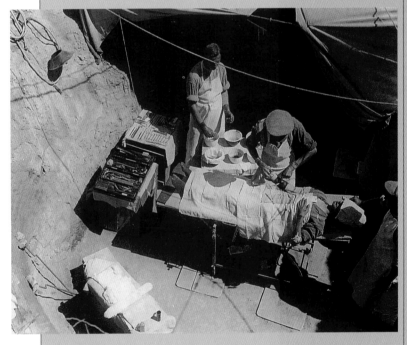

FIELD HOSPITAL A surgeon removes a bullet from a man's arm. The "dug-out" operating room gave some protection from snipers and shells.

appalling casualty rates, with one in four soldiers dying from their wounds in 1915. One of the reasons for this was that when casualties were operated on, dead or infected tissue was not always fully removed—or débrided—before wounds were sewn up. This often

All forces involved in the Second World War had access to sulpha drugs, but penicillin was available only to the Allies. This valuable drug arrived more than halfway through the conflict, in 1943, and in that year was available only in tiny quantities. By 1944, however, production had leaped upward, with more than 3 billion doses being produced. Soon after the end of the war it was being used worldwide.

During the Korean War (1950-3), American and United Nations forces were involved in warfare over mountainous terrain, where rapidly moving front lines eventually gave way to stalemate. In the American forces, this conflict saw the introduction of Mobile Army Surgical Hospitals—MASH for short—which were subsequently immortalized on television and in film. The aim of these units was to bring high-level medical care as close to the battlefront as possible, so that severe injuries could be treated with the minimum of delay. Each unit had up to 200 beds and a total of 60 medical and nursing staff, and could deal with several thousand patients a month. Well-equipped operating rooms allowed surgeons to carry out complex operations, including repair work on severed blood vessels, a procedure normally undertaken only in hospitals.

The Korean conflict saw the first large-scale use of helicopters in medical emergencies. In the final years of the Vietnam War, which lasted from 1965 to 1975, helicopters became ambulances of the air. While Vietcong casualties were passed slowly back down long lines of supply, American and South Vietnamese soldiers were often flown out of the area of conflict, with blood transfusions being given en route. This kind of rapid evacuation had beneficial effects: in the Second World War, one in four wounded American soldiers died from their wounds; in Vietnam the figure was lower than one in six.

In the Gulf War in the 1990s preventative treatment was seen as an important defense against new forms of weaponry. British and American troops were given drug treatments to off-set the anticipated effects of chemical and biological weapons—with controversial results. The claimed side effects of this treatment—known collectively as Gulf War Syndrome—remain a matter of medical and political debate.

HIGH-TECH WARFARE A helicopter takes away wounded in Vietnam, while a Gulf War soldier checks his gas mask.

DOUBLE-DECKER AMBULANCE Injured American soldiers are flown out of Korea in 1952 on board a giant C-124 troop carrier.

THE SECOND DEADLY DOSE

At the beginning of the 20th century, one kind of medical emergency seemed to elude any rational explanation. A seemingly healthy person would suddenly have difficulty breathing, and by the time help was at hand they might have collapsed, and even lost consciousness. The trigger for this dramatic chain of events

would be something apparently quite trivial—such as an insect sting—an event that, in someone else, would produce no more than passing discomfort.

In the course of research into immunization, Charles Richet, a French professor of physiology, started to throw some light on this kind of overreaction. During 1902 and 1903, he isolated the proteins from jellyfish stings, and then injected tiny amounts into a dog. The dog showed no ill effects after the first dose, but when Richet gave it a second injection a few days later, it promptly died.

Richet came up with a new term for this bizarre effect: he called it anaphylaxis, in contrast to prophylaxis, which means the prevention of disease. During anaphylaxis, the body's immune system goes into a form of defensive overdrive, and the result is a condition that, far from helping, can itself be life-threatening.

Today, reactions of this type are known by the blanket term allergy (from the Greek *allos* meaning "changed" and *ergos* meaning "action"). Coined in 1905 by Clemens von Pirquet, an Austrian pediatrician, it provides a convenient label for a range of medical problems caused when the body overreacts.

In the course of the century, allergies have been the subject of many kinds of treatment. These include antihistamine drugs, which were developed in the 1940s, and immunotherapy treatment, which involves giving the sufferer gradually increasing doses of the triggering substance—or allergen—so that the immune system becomes desensitized. However, for many allergy sufferers, the only effective treatment is to avoid the substance that causes the problem—which is often easier said than done.

ALLERGY EXPERT Charles Richet's work on allergies and anaphylaxis earned him the Nobel prize for medicine in 1913.

often start to beat normally once more.

In 1947, an American surgeon, Claude Beck, became the first person to try this resuscitation technique, called defibrillation, on a human patient. During an operation on a 14-year-old boy, the boy's heart stopped. After an hour and a quarter of fruitless direct massage, he decided to try an electric current. When the electrodes were applied to the boy's heart and the current switched on, the heart was jolted back into life.

Many cases of fibrillation or cardiac arrest happen away from hospitals, where rapid surgery is not an option. During the late 1940s and early 1950s, a number of medical teams worked to devise an external version of the defibrillator, which would kick-start the heart from outside the body. Their efforts bore fruit, and the external defibrillator was first successfully used in 1955.

Dealing with shock

One of the most common effects of any severe injury is shock. Used in its medical sense, the term does not mean mental dis-

MECHANICAL AIDS Bernard Draeger's Pulmotor (bottom) paved the way for the medical ventilator. The portable defibrillator made its first appearance in the mid-1950s.

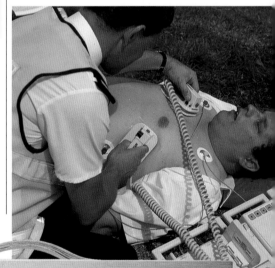

automatically. Called the Pulmotor, it had a clockwork motor that controlled the movement of oxygen from a pressurized cylinder. Once the lungs had been inflated with oxygen, the Pulmotor reversed the pressure in the supply, so the patient's lungs slowly deflated. The motor would then move on, and the cycle would begin again. The Pulmotor met with a mixed reception. However, it showed that machines could be used to assist breathing and became the forerunner of the clinical ventilator, a lifesaving device that is now used in hospitals throughout the world.

The decade that saw the birth of the Pulmotor also witnessed a step forward of a dramatic and quite different kind. In 1901, during an operation for a cancer of the uterus, a Norwegian woman underwent cardiac arrest, a situation that at the time was normally fatal. The surgeon—Dr. Kristian Igelrud—quickly made an incision in the

patient's chest and massaged her heart by hand. The woman's heart started beating again, and she eventually recovered.

In the 1930s, researchers discovered that small electric shocks could make the hearts of animals fibrillate, a state of rapid and erratic beating that often precedes cardiac arrest. But if a much more powerful shock was given, the heart's normal pacemaking system would be reset, and the heart would

RAPID RESPONSE

IMPROVEMENTS IN TRANSPORT, COMMUNICATION AND PORTABLE MEDICAL EQUIPMENT HAVE OPENED UP NEW HORIZONS IN ON-THE-SPOT EMERGENCY CARE

From the early days of motorized ambulances at the beginning of the century, paramedics—medical auxiliaries who specialize in emergency treatment—had to be familiar with the problems of transporting the injured. In Europe, where many roads were still surfaced with cobblestones, the ride to the nearest hospital was often a bumpy and potentially dangerous one. Improved vehicle suspension helped, but patients still had to be carefully strapped to a stretcher to prevent sudden jolts having disastrous results.

AT THE FRONT LINE This First World War motorcycle ambulance had removable stretchers for two casualties.

Before the 1930s, the equipment carried on ambulances was relatively basic, and paramedics needed no specialized knowledge beyond a good understanding of first aid. But by the time the Second World War was over, on-the-spot treatment had become much more sophisticated. In the late 1940s, ambulance crews routinely set up emergency blood transfusions and intravenous saline drips. Portable oxygen cylinders and manually operated ventilators were often carried on board.

Since the 1940s, improvements in technology have allowed a steadily increasing range of treatments to be carried out in the crucial minutes immediately after accidents and other medical emergencies. Disposable syringes now allow paramedics to administer drugs in fast and accurate doses. These drugs include painkillers and other substances such as adrenalin, which can make the difference between life and death in cases of heart failure and severe allergic reaction. The portable defibrillator, which found its way onto ambulances during the early 1960s, has also become a key item of equipment, enabling a patient's heart to be restarted before oxygen shortage creates permanent damage. And radio communication between paramedics and hospital emergency rooms means that by the time accident victims arrive, an initial assessment has been made and further treatment can be started without delay.

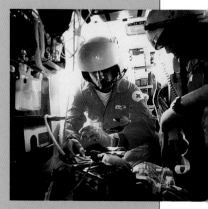

AIRLIFT A German helicopter ambulance team tends a road accident victim.

MODERN AMBULANCE Today's ambulances provide high-tech care within minutes of an emergency call.

Primary response pack	Portable oxygen supply and ventilator	Cardiac monitor and defibrillator
Kit bag		Tools
		Orthopedic stretcher
		Pulse oximeter
Safety helmets	Aspirator (suction unit)	Entonox set (pain-relieving gas)
		Rechargeable handlamp
Paramedic kit bag	Spinal board	Various equipment cases
Oxygen masks	Blanket	

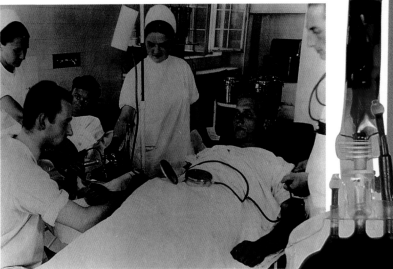

tress. It refers to a potentially catastrophic reduction in blood pressure—one brought about either by extensive bleeding, or through a sudden failure of the circulatory system. Left untreated, shock can be fatal.

For a doctor attending an accident victim in the early part of the century, shock was a formidable adversary. Over 150 years ago, medical research showed that shock could sometimes be treated by injecting saline solution into the patient, helping to return the body's fluid balance to nearer its normal level. However, it was not until the 1930s that intravenous saline infusions—initially from bottles, and then from plastic sachets—became a routine part of on-the-spot treatment.

A saline drip can be safely administered to almost anyone, but a more obvious way of combatting shock—by replacing lost blood—poses far greater problems. Early experiments often had fatal results, and by 1900, blood transfusions had acquired a dangerous reputation. In many countries they were

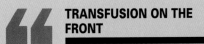

GIVING BLOOD In the 1930s, transfused blood often came direct from the donor. Today, all blood is screened and prepacked (right).

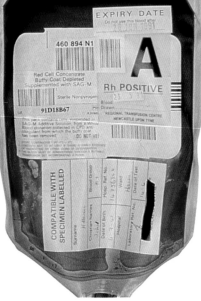

banned. This was the year in which Karl Landsteiner, an Austrian pathologist, started to investigate the difficulties of transfusing blood. Sometimes transfusions succeeded, but often—for no apparent reason—the recipient reacted very badly, and many died. Landsteiner set out to uncover some logic behind this perplexing pattern.

As a first step, he took small amounts of blood from himself and a group of colleagues. He then mixed blood samples from each person with samples of serum from the others (serum is the part of blood remaining when its cells have been removed). The outcome was decisive. Some of the mixtures remained fluid, but in others the red blood cells clumped together, or agglutinated. Landsteiner correctly reasoned that this clumping was triggered by some kind of immune reaction, and that it accounted for the symptoms seen when blood transfusions went wrong. In fact, the clumping together of red blood cells causes blockages in blood vessels, often with fatal results.

Once Landsteiner had discovered agglutination, he set about identifying its precise causes. He found that some people's red blood cells carry a surface protein that he labeled A, and others have a different protein, which he labeled B. Some people have both, and some have neither. This gives four possible combinations, or blood groups: A, B, AB and O. He also found that the fluid part of the blood could contain two other proteins, called anti-A and anti-B, which can react with the surface proteins to cause clumping. By testing all the possible combi-

nations of donor and recipient blood, he established which blood types were compatible with each other, and therefore which transfusions were safe.

A successful blood transfusion was performed in 1905, and by 1910 it seemed likely that transfusions would become routine. However, in these early transfusions, the blood was transferred directly from donor to recipient while still warm. This system was necessary because blood soon clots when it is removed from a donor's body. Before blood could be stored, a way was needed to prevent clotting without altering the blood's other characteristics. Several chemical additives were tried, until one—sodium citrate—showed itself to be safe. In November 1914, just after the outbreak of the First World War, the first transfusion took place with citrated blood. By 1935, the first blood banks were set up, and since then, blood transfusions have saved millions of lives.

TRANSFUSION ON THE FRONT

At the height of the Spanish Civil War, the British physiologist, geneticist and writer J.B.S. Haldane was helping to provide medical assistance for the Republican cause. As General Franco's Nationalist forces laid siege to Madrid in 1937, Haldane worked alongside Dr. Norman Bethune, who later became famous for his work in China. Haldane wrote many newspaper articles on developments in medicine and science, and in one he described witnessing the life-giving power of a blood transfusion firsthand:

"A successful transfusion is a wonderful sight. In January 1937, I accompanied Dr . Bethune, a Canadian who started the first transfusion unit in Madrid, to a field hospital north of the town. A Spanish comrade was brought in with his left arm shattered. He was as pale as a corpse. He could not move or speak. We looked for a vein in his arms, but his veins were empty. Bethune cut through the skin inside his right elbow, found a vein, and placed a hollow needle in it. He did not move. For some twenty minutes I held a reservoir of blood, connected to the needle by a rubber tube, at the right height to give a steady flow. As the new blood entered his vessels, his color gradually returned, and with it consciousness. When we sewed up the hole in his arm he winced. He was still too weak to speak, but as we left him he bent his right arm and gave us the Red Front salute."

METHODS OF TREATMENT

BEFORE 1900, DOCTORS OFTEN HAD LITTLE MORE TO OFFER THAN COMFORTING WORDS AND LARGELY INEFFECTIVE REMEDIES. SINCE THEN, A HOST OF BREAKTHROUGHS—FROM VACCINES TO ORGAN TRANSPLANTS—HAVE OVERCOME MANY DISEASES AND DISORDERS THAT ONCE PROVED FATAL. DESPITE THESE SUCCESSES, MODERN LIVING AND MEDICAL TREATMENT HAVE BROUGHT NEW HAZARDS AND DILEMMAS, PERSUADING MORE AND MORE PEOPLE TO TRY ALTERNATIVE TREATMENTS.

FROM HOME TO HOSPITAL

DURING THE 20TH CENTURY, HOSPITALS DEVELOPED INTO CENTERS OF SPECIALIZED KNOWLEDGE AND THE FOCAL POINT OF TREATMENT

Hospitals are institutions with a long and varied history. Originally places of rest for the needy and infirm, they gradually opened their doors to a much wider section of the public. In 1656, for example, the new Hôpital Général (general hospital) in Paris not only provided a refuge for the old and a prison for the insane, it also doubled as a place where healthy people, including troublemakers, idlers and prostitutes, could be kept off the streets. This rambling institution housed more than 6,000 people, and while it helped to promote public order, it did little to treat disease.

Today, hospitals are run with very different

their lives, but when the century opened, being a hospital patient was a rare event. Medical specialization had not reached its modern extent, and hospitals were places where the long-term sick were nursed, and where only complex treatments—such as major surgery—were carried out.

The move from home to hospital treatment began slowly, but by 1900 was well under way. Two decades before the new century began, New York Hospital announced that its operat-

ing room facilities would be superior to those available in the "most luxurious home"— a reference to the days, not long past, when most operations were performed in domestic surroundings. But by the time the new century arrived and operating rooms were in use, few people needed to be persuaded that, as far as surgery was concerned, a hospital was far safer than home.

When the Paris Hôpital Général was built, little

MEASURED DOSE A nurse pours out a dose of medicine. In 1900, most medicines came in liquid form and were measured out in glass minim measures (far right).

aims. But even in the 20th century, the hospital has continued to evolve. By the 1990s, most people in developed countries experienced at least one stay in a hospital at some point in

thought was given to designing it to suit medical needs. But after Louis Pasteur's discoveries in the mid 19th century about the role of bacteria in causing disease, hospital design changed radically. By the time the 20th century began, every detail of new hospital buildings had to prove its medical worth. In a hospital design handbook of the 1920s, the American architect Edward F. Stevens showed that this process had begun. "Health values do not reside exclusively in smooth walls, smooth floors, and rounded

1913 *The Modern Hospital* handbook on hospital design

1920s The Scialytique shadowless light invented for operating rooms

1930s Modern multi-positional operating table developed
In-house hospital training becomes routine for doctors
Hospital beds with wheels and adjustable backrests

1940s Air conditioning introduced into operating rooms

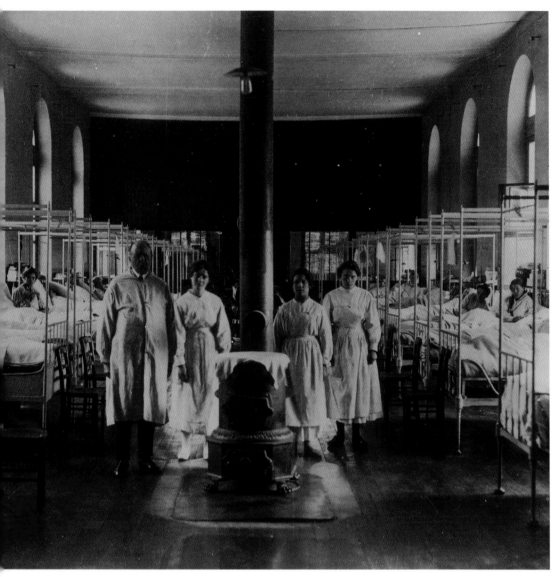

PATIENTS ON PARADE A French hospital ward in the 1930s, with huge windows and neat rows of beds, exemplifies hygiene and uniformity.

disposables such as syringes, and most hospital drugs—including powerful narcotics—were still kept on open shelves.

Life on the ward

In the early decades of the 20th century, hospital wards were almost always designed around a single central aisle, flanked by two rows of beds. The nursing station was located at the head of the aisle, allowing the staff to see all the beds at a glance, and behind it, through a storage and utility area, was the corridor leading to the rest of the hospital. A typical hospital ward might hold up to 40 patients, and as wards were open-plan, privacy was limited. Beds could be screened by wooden or metal frames, but these had to be moved about the ward as necessary. These pieces of equipment did not have a

INTENSIVE CARE Nurses use modern technology to monitor patients at the University of Pennsylvania.

inner corners," he wrote, citing some recent improvements designed to promote hygiene. He went on to list many other features that should be considered or incorporated when a hospital was being planned. These included properly oriented wards, balconies positioned to catch the sunshine, grounds or flat roofs that were accessible to patients, sanitary construction, proper sleeping quarters for resident employees, devices for noise prevention and, above all, a "cheerful and tonic outlook." The message was clear: the modern hospital should strike an optimistic note, helping to speed the patient's recovery.

Despite this movement toward modernity, hospitals in the 1920s still lacked many of the facilities that we take for granted today. There were no intensive-care units, often no postoperative recovery rooms, and no electrical resuscitators for reviving patients who had suffered cardiac arrest. There were no portable paging systems, no

1960s Hospital beds with height adjusters

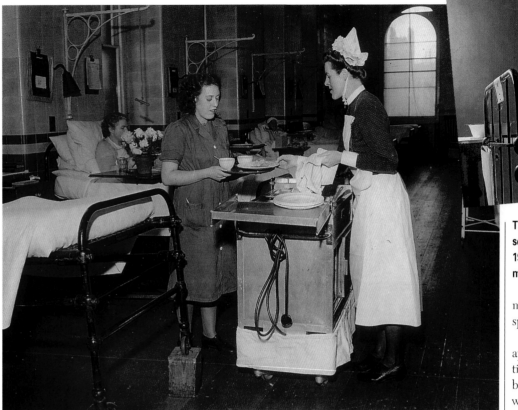

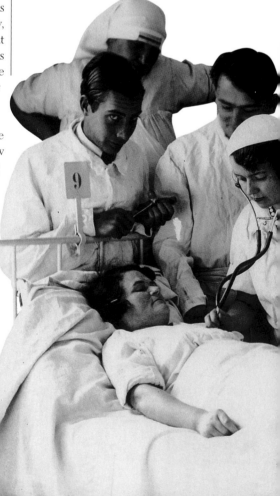

THE DAILY ROUTINE A nurse and assistant serve up lunch in a British hospital in the 1940s (left). In an American hospital, a nurse methodically cleans an empty bed (above).

good reputation. An acerbic paragraph printed in *The Modern Hospital*, published in 1913, echoed views that must have been held by countless nurses at the time: "There are many such screens on the market, and none of them is acceptable . . . If they are made of metal they are too heavy to be moved readily, especially when the material is of heavy cloth or canvas, and where they have castors, the castors are usually out of order. If they are made of wood and are

---

IN BED OR OUT

Until the 1980s, hospital life was strongly influenced by the idea that complete rest was an important aid to recovery. As a result, physical activity was discouraged, and patients often spent long periods confined in bed. Today, medical thinking favors exactly the opposite approach. Many patients are encouraged to get on their feet as soon as possible, to prevent the muscular and circulatory problems that come with a prolonged stay in bed.

---

light, the slightest push knocks them over, and the paint is soon worked off by cleaning, and they are not very sanitary." It is hard to imagine why curtains took so long to appear.

The task of keeping a ward clean and tidy was a major preoccupation for nursing and auxiliary staff, and during the first half of the century—when military uniformity was still

deemed essential in hospitals—it could border on the obsessive. One British patient, writing in the 1930s, noted the great amount of time that was spent polishing ward floors. In an era when flooring was made of wood rather than plastic, a mirror-like sheen was synonymous with hygiene; unfortunately, newly polished floors were so slippery that they could be difficult and even dangerous to walk on. In hospital wards throughout the United States and Europe, a similarly dogmatic approach was often applied to bedding. Every patient had his or her bed made up in the same way, with the same number of pillows, regardless of how comfortable or uncomfortable the bedding made them feel.

There were some pluses. The open nature of these wards often evoked a spirit of camaraderie, with patients encouraging each other. However, in some circumstances openness was a disadvantage. Before the century began, death was still a relatively public event in the ordinary ward, but by the 1920s there was growing recognition that death— and the grief that inevitably surrounds it—should not be exposed to the curious eyes of other patients and their visitors. As a result, the most severely ill were

moved into side wards, where they could spend their final days in privacy and peace.

After the Second World War, a new approach to ward design made the segregation of patients easier than before. Instead of being arranged around a single corridor, wards in many newly built hospitals now had a double corridor that ran down one side of the ward, and then back along the other. The space between the corridors was taken up by the nursing station and utility areas, with the patients' beds arranged on either side.

Frequent rearrangement of the wards was helped partly by developments in the hospital bed. In the Victorian era, some beds had wheels, but most were designed like their domestic equivalent, and made to stay in place. In the early decades of the 20th century, the hospital bed was still largely immobile, although small wheels were fitted on some. Most were made of metal covered with white enamel, a finish that made dust more noticeable and easy to remove.

In 1913, the author of *The Modern Hospital*, Dr. John A. Hornsby, deplored the tendency to complicate the hospital bed by "the addition of all sorts of alleged conveniences." But such views did not prevail. By the mid-1930s, the formerly humble bed had become much more sophisticated, with large wheels and built-in backrests. And in the 1960s, beds with height and posture adjusters started to appear. The inflatable mattress also arrived, helping long-term patients avoid bedsores.

SHOT IN THE ARM Until the 1960s, reusable syringes were standard equipment. This pocket version from the 1930s has its own metal case.

Divided into separate compartments that are inflated and deflated in turn, this kind of mattress alternates the parts of the body in contact with the bed at any one time.

The ward round

Despite changes in ward design, one aspect of ward life has remained a fixture since the century began. This is the daily ward round, in which a consultant or doctor assesses the

CENTER OF ATTENTION A patient undergoes examination surrounded by a bevy of doctors and nurses in the early 1930s.

THE MISSING WHITE COATS

With the growth in services provided by hospitals, clothing plays an important part in distinguishing between different types of healthcare professional. However, as the American sociologist Julius Roth noted in 1957, what hospital staff do not wear often says as much about them as what they do.

Roth was investigating staff in a hospital treating victims of tuberculosis, where gowns, masks and hair coverings were meant to be worn by all staff coming into contact with patients. He found that students wore the obligatory protective clothing every time they entered the room that Roth was monitoring. Assistant nurses wore their gowns and masks about 50 percent of the time, while the more senior nurses wore them on less than one occasion in five. Doctors hardly wore protective clothing at all.

Roth attributed these differences to status and prior knowledge. Not wearing a uniform was—and still is—a sign of status, suggesting a rank that transcends the rules. Students, on the other hand, follow the rules to emphasize their inclusion in the medical world. However, Roth also discovered that most senior staff knew protective clothing had little effect.

state of each patient and directs the treatment given by the staff of nurses. The public nature of the original single-corridor wards gave the round a flavor that, even in the more compact wards of today, has not been entirely lost. In the 1920s, an age of deference to figures of authority, the consultant reigned supreme and his progress down the ward would be greeted with respectful silence (at the time, consultants were invariably male). No one, least of all the patient, would dream of questioning his authority or asking for explanations of the treatment prescribed.

Despite the persistence of this daily routine, some features of the round have changed. By the 1960s, an era when authority of all kinds was being openly questioned, the formality of the round began to wane. Another change was that the ward round had now become a teaching instrument—a vital part of training for students in the early stages of a medical career.

From a modern point of view, it seems obvious that medical students should learn their skills by watching experienced physicians at work. But it was not always like this. During the late 19th century, selected patients were brought from hospital wards into medical school lecture halls as exhibits, and many doctors embarked on their careers without ever once setting foot in a hospital. Even as late as 1914, American medical students could qualify without hospital experience; that often came only after they had completed their degree. But by the 1930s, in-house training had become a required part of a medical career.

Changing rooms

To a patient from the early 20th century, some features of today's hospital wards would still be recognizable, while in other departments, much would be strange and unfamiliar. Whole suites of rooms are now given over to diagnostic machinery—such as

ating room, a facility that now dominates hospital life.

During the early decades of the century, surgery underwent explosive growth. In many European hospitals, the number of operations carried out between 1900 and 1930 more than tripled, and in some North American hospitals they increased by a factor of five. At the Pennsylvania Hospital, for example, the most commonly performed operation—removal of the tonsils—was carried out about two dozen times in 1895, but more than 1,000 times annually 30 years later. As operations became more complex,

time, many surgeons still worked by natural light, although spotlights were often mounted above the operating table itself. Many surgeons disliked these overhead lights, however, because as soon as they leaned forward to make an incision or sew up a wound, they would cast a shadow over the patient. To get around this problem, spotlights were often arranged on a wide frame or ring, so that one lamp would always fill in the shadow thrown by another. Later in the 1920s, a further improvement came in the form of a single shadowless light called the Scialytique, which was developed by a French professor of surgery, L. Verain. The Scialytique consisted of a powerful bulb set in the middle of a saucer-like reflector, with a mask in front of the bulb to prevent light from shining directly on the table below. A circle of light was reflected from the saucer's rim, so that it converged on a narrow spot. This new light could be trained on any part of the operating table, much like operating room lights today.

As artificial light improved, surgeons retreated into an interior world where surgical teams could work at all hours of the day and night. By the 1940s, the arrival of air conditioning meant that most operating room windows were kept permanently shut, and by the early 1960s, operating rooms were being built without any windows at all. As a result, the surgical suite lost its elevated position in the hospital building, and began to be located in the building's core.

In the early days of hospital surgery, each surgeon would have a personal set of instruments. By the time the 20th century got under way, the era of personal instruments was already past. Improved manufacturing techniques meant that one standardized

HEALTH HAZARD An operation in the early 1900s is watched by visiting doctors. Despite the danger of infecting the patient, the onlookers are wearing everyday clothes.

MRI (magnetic resonance imaging) scanners—which did not exist even 30 years ago, and complex technology pervades hospitals on all fronts, from the intensive-care units to the kitchens. Some of the most far-reaching changes of all have come about in the oper-

and more staff and equipment had to be crowded into the operating rooms, the spaces themselves started to expand. In the 1930s, a fully equipped operating room might have a floor area of about 300 square feet, but by the time heart surgery became feasible in the 1960s, hospital operating rooms could be double that size.

In the 1920s, hospitals often had their surgical suites on an upper floor, where they were farthest from dust in the street. At this

instrument was much like the next, and with the development of stainless steel just before the First World War, all surgical instruments could be routinely autoclaved (sterilized by high-pressure steam), instead of having to be disinfected. By the 1950s these instruments had become just a small part of the surgeon's basic needs. More ambitious surgery demanded complex technical backup, including air lines to drive drills and circular saws, oxygen for ventilating machines, and electrical monitors for checking the patient's pulse and blood pressure.

Like the hospital bed, the operating table also changed. Before the 1920s, most operating tables resembled trolleys, with a platform that could be tilted into different positions.

PRECISION ENGINEERING This Swiss-made operating table dates from the mid-1930s. It can be adjusted for use in more than 20 different positions.

These included a position that raised the pelvis for operations on the kidneys, a sideways sitting position, used for some operations on the chest, and the "Trendelenburg position," in which the patient's body sloped head downward to keep blood out of the lungs. Although the wheels could be locked, tables like these were prone to wobble, and the surgeon often had to lean over the patient to reach the site of the operation. However, within two decades the operating table had been completely rethought. The trolley-like legs disappeared, and the base became a massive flattened tripod—heavy enough to keep the patient still but compact enough to be out of the surgeon's way. The table itself was supported on a single telescopic pillar, which could be raised or lowered in seconds, and it could be fitted with an electric heater to keep the patient warm. Operating tables like this, which were available in 1935, are not much different from those in use today.

Special hospitals

Alongside general hospitals dealing with all types of disorders, there were specialized institutions tailor-made for patients who had much more specific needs.

CONTEMPORARY FACILITY An open-heart operation in progress at an American hospital. Unlike earlier operating rooms, today's are practically sterile environments.

As the 20th century began, tuberculosis was in decline, but it was still one of the most widespread and feared infectious diseases. In the absence of effective drugs, the only treatment was a prolonged stay in a sanatorium. Unlike other hospitals, sanatoriums dealt exclusively with tuberculosis patients. Their main weapons in the fight against the disease were gentle activity and rest, together with sunshine and fresh air.

Sanatoriums varied widely between one country and another, and according to the patient's social class. In Europe, the wealthi-

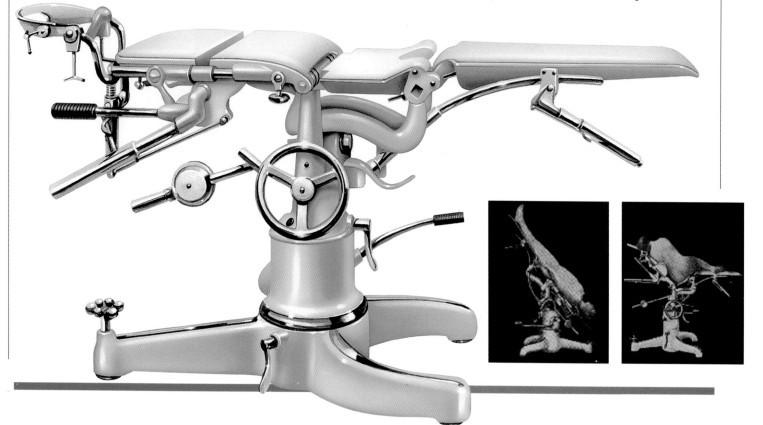

ON SCREEN, ON CALL

FROM "DR. KILDARE" TO "ER," TV DOCTORS HAVE BROUGHT
MEDICINE INTO AMERICAN HOMES FOR DECADES

In the early days of television, the "doctor show" was a programming staple. Doctors became a regular presence as television expanded into homes across the nation. One show, "Medic," from the mid-1950s, focused on the work of a doctor played by Richard Boone. The show aimed to portray the profession realistically, and consistently emphasized the doctor's skill and assurance. Boone played the role with a lack of emotion and a detachment thought appropriate for the work of a doctor.

Later shows such as "Ben Casey" and "Dr. Kildare," both from the 1960s, portrayed doctors in a more personal light, depicting them not only as professionals but as multi-sided characters. "Dr. Kildare" concerned the relationship between an intern and an older physician, who taught the student (and the audience, by extension) about what it means to be a doctor. Likewise, "Ben Casey" was about a neurosurgeon who relied on the advice of his older mentor.

In both "Dr. Kildare" and "Ben Casey," episodes featured doctors attempting to solve medical problems. The young doctor would have to rely on his knowledge and growing experience to make a decision—one that may or may not have been correct to save the life of the patient. The shows provided a serious, if not always realistic, look at how doctors practiced medicine, and at the ways in which difficult medical decisions were made.

But in the 1970s, the portrayal of doctors on television took a comic turn with the debut of "M°A°S°H," a show about military surgeons that captured the American imagination for more than a decade. In fact, some say "M°A°S°H" was the best show of all time.

The series chronicled the experiences of surgeons operating in the 4077th Mobile Army Surgical Hospital in Korea. Although "M°A°S°H" dealt with larger issues, including the absurdity and insanity of war, it remained at its core about the work of doctors, which gave the show a medical feel.

Television doctors returned to the hospital setting in the mid-1980s, when Americans in droves tuned in to the doctors and nurses on "St. Elsewhere." Set in a rundown Boston neighborhood, the drama combined realism with dark humor in its depiction of efforts to provide care to those not served at more prestigious facilities. On the show, doctors' work was interrupted by muggings in the emergency room, nurses stealing from patients and other distractions. The show won praise for its realistic look inside a hospital environment, with hand-held cameras and overlapping dialogue used to increase the sense of authenticity.

"St. Elsewhere" went off the air in 1988; six years later, medical drama returned to the airwaves in the hit series "ER." Set in a fast-paced Chicago hospital, "ER" was based on a script by best-selling author Michael Crichton. The show surpassed "St. Elsewhere" in its degree of realism, relying on graphic emergency room scenes, and with actors using actual medical terminology and making life-and-death decisions as quickly as they must in real life.

Like any television show, "ER" has had its share of romance and drama; but more so than any of its predecessors, "ER" has succeeded in showing viewers what life is like for doctors and nurses in a big-city hospital. Viewers have responded to this realism and intensity, making "ER" one of television's most consistent hits throughout the late 1990s and into the new century.

AMERICA'S DOCTOR Richard Chamberlain portrayed TV's "Dr. Kildare" in the 1960s.

ANOTHER COUNTRY A scene from the TV series "M*A*S*H," set in the 4077th Mobile Army Surgical Hospital in Korea.

est tuberculosis victims could take refuge in private sanatoriums in the Swiss or German Alps, where services were often similar to those of first-class hotels. Some were equipped with rotating solariums designed to expose patients to the maximum amount of beneficial sunshine. For those of more modest means, a period in a sanatorium often

CLASS APART The treatment for tuberculosis patients at a Swiss sanatorium in 1911 (right) centered on rest and fresh air. Inmates of a German psychiatric hospital in the 1930s (below) are encouraged to exercise.

meant a stay in less luxurious surroundings closer to home. One of the largest British sanatoriums, the King Edward VII, opened in 1906, typified the kind of regime then believed to accelerate a cure. The patients were woken at 7:30 a.m., and from then until 10 p.m., when lights were turned out, every moment was accounted for in the rigid timetable. This included reading from 9 to 9:30 am. (fast-moving novels and books about tuberculosis were not permitted), exercise from 10 to 12, handicrafts between 2 and 2:30, and periods of enforced rest after meals. Many patients lived by this timetable week after week for several years.

With the introduction of antibiotics and vaccines, patient numbers dropped and the sanatorium system declined. But during

UNDER RESTRAINT An American psychiatric patient in the late 1940s is kept strapped to her bed.

the same period, another kind of medical institution was struggling to cope with rising numbers. By 1900, the mental asylum—literally, a place of refuge for the mentally ill—already had a long and dubious history, but in the decades before the 1950s the number of patients in asylum hospitals increased rapidly. At the beginning of the century asylums in the United States held about 150,000 inmates; just five decades later they housed over half a million. Similar growth in patient numbers took place in Europe: by 1916 Italy, for example, already had 40,000 inmates, while Britain in the 1950s had 150,000. While many inmates were indeed ill,

many others—particularly in the 1920s and 1930s—were admitted for what would now be considered social problems rather than mental ill health.

As the decades went by, it became clear that hospitalization itself was not a cure. By the 1970s, psychotherapy and treatment with drugs had begun to replace confinement, and slowly the days of the giant asylum came to a close. Some of these cavernous institutions still remain, but most—like the sanatorium—have slowly slipped into history, leaving few reminders of the lives that they once constrained.

FIGHTING INFECTION

SINCE THE LATE 19TH CENTURY, THE SCIENCE OF IMMUNOLOGY HAS CURBED DISEASES THAT ONCE KILLED MILLIONS EVERY YEAR

In one of the most popular paintings of the late 19th century, the British artist Sir Luke Fildes depicted a poignant scene that would have been familiar to many. In the dim light of a darkened room, a doctor thoughtfully clasps his chin, while on a bed nearby, a child lies in the grip of a potentially fatal illness. Copies of the painting found their way into homes and doctors' offices throughout the world, and it was reputed to have been reproduced over a million times. The reason for its appeal was simple: it represented a shared moment of crisis that only skill and care could resolve.

Just over a century later, fatal childhood infections, along with many that once afflicted people in later life, are largely a thing of the past, although the agents of infection still abound. This achievement has been brought about through three lines of attack. The simplest of these has been the improvement in public hygiene, which was well under way by the time the 20th century began. A second line of attack—drug therapy—had its first success in the century's opening years, but only came into prominence several decades later. At the time Fildes painted *The Doctor*, and during the years that immediately followed it, the fight against infection was focused in a new and rapidly growing science: immunology.

Biological defenses

By the beginning of the 20th century, bacteria had been linked to infectious diseases, and the body's key defense mechanism, the immune system, was under growing scrutiny. Immunity is the capability of the body to disable or destroy invading microorganisms. Some diseases, such as the common cold, are soon fought off but leave us vulnerable to future attack. Many others—including chicken pox and measles—trigger acquired immunity, which ensures that we never catch the same disease again. Knowledge of this valuable form of immunity is not new. As long ago as the 1700s, people discovered by accident that in the case of smallpox, acquired immunity could be produced by artificial means. This originally involved scratching infectious pus—initially from a smallpox pustule, but after Edward Jenner's work in the 1790s, from a cowpox pustule—into the skin of an uninfected person. Called vaccination, this procedure eventually became widespread, although no one knew how it worked.

In the late 19th century, when more was known about the causes of disease, French and German researchers tried to find out how the immune system operated, and how it could be exploited. At the Pasteur Institute in Paris, the Russian-French zoologist Elie Metchnikoff advanced a theory based on cells. He had seen how simple forms of life, such as amoebas, feed by engulfing microscopic particles of matter, and later observed that white blood cells do the same thing with bacteria. From this, he developed the cellular theory of immunity—the idea that the body's defensive cells track down invading microorganisms and engulf them to prevent them taking hold.

Across the border in Germany, a different view prevailed. Here Robert Koch, Professor of Public Health in Berlin and founder of medical bacteriology, was inclined to dismiss the role of defensive cells. It had been shown that serum—blood with its cells removed—could kill bacteria if it came from an animal that was already immune. Even more significantly, the same serum would continue to work if it was injected into another animal's blood. Clearly, something dissolved in the serum was producing an immune response. The defensive substances were named *Antikörper*, "antibodies." For the German researchers, these formed the central feature of the humoral theory of

BATH TIME Cheerful smiles mask the fact that, in 1901, babies stood a significant chance of dying from childhood disease.

immunity, so-called after "humors," an archaic name for the body's fluids.

In the decades that followed, both theories were proved partly correct because the immune system depends both on cells and on antibodies that circulate in the blood. But despite this disagreement, the study of immunology paid early dividends. By developing techniques to produce immunity artificially, several major killers were rapidly brought under control.

The end of diphtheria

At the beginning of the 20th century, immunologists already knew that the process of inducing immunity could be dangerous. If a vaccine contained fully active bacteria, there was a chance that it could trigger the disease that it was designed to prevent. One way around this problem, discovered by the French bacteriologist Louis Pasteur in 1879, was to weaken the organisms that cause disease. Pasteur found that if bacteria were grown in the laboratory for a significant length of time, they would often lose their virulence while retaining their power to provoke immunity. Pasteur managed to strike this balance with anthrax, a disease found in cattle and to a lesser extent in people, but other diseases proved to be more problematic.

In Germany, Emil von Behring, one of Robert Koch's colleagues, tried another approach. He was investigating diphtheria, a respiratory disease that was then one of the

TETANUS PREVENTION Emil von Behring's tetanus antitoxin, produced in the 1890s, made immunization possible.

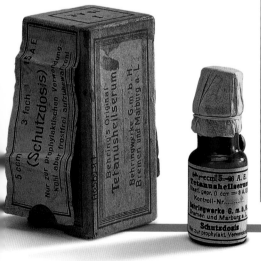

ONE AT A TIME Dr. Bela Schick administers a diphtheria test to New York schoolchildren in 1925 (above). Advertisements in doctors' offices, such as this one from Britain (right), also played a part in ensuring that children were immunized.

most dangerous childhood killers. Diphtheria infects the mouth and throat, and forms a characteristic gray membrane that spreads down the windpipe. This alone can cause asphyxiation, which at one time could be countered only by a tracheotomy—an emergency incision opened up in the throat. But the bacterium also attacks the body by releasing a toxin so powerful that just two-millionths of an ounce can kill a child.

Instead of preparing a vaccine directly, Koch first injected bacteria into horses, and then injected a second batch of animals with small amounts of serum taken from the first. These horses did not develop the disease, but they did form defensive chemicals, called antitoxins, to counter the bacteria's deadly poison. When Behring then injected antitoxin into a sick child, the effect was remarkable and almost immediate: the child recovered. This experiment, which followed similar

DIPHTHERIA is deadly-

IMMUNISATION is the safeguard

ASK AT YOUR LOCAL COUNCIL OFFICES OR WELFARE CENTRE

Issued by the Ministry of Health and the Central Council for Health Education

work with tetanus, resulted in thousands of children surviving a formerly fatal disease. However, although Behring's treatment cut the death rate from the disease, it did not prevent it.

The final breakthrough against diphtheria was made by the French bacteriologist

Gaston Ramon in 1923. He found that the toxins made by diphtheria and tetanus bacteria could be altered by heat and by treating them with chemicals. The result, called a toxoid, proved harmless but produced long-lasting immunity. With mass immunization of children from the 1930s on, diphtheria practically disappeared.

The BCG controversy

In the history of 20th-century immunology, few vaccines have aroused such furious and protracted debate as one known as Bacillus Calmette-Guérin, or BCG. Developed in 1908 to combat tuberculosis, it was the work of the French bacteriologist Albert Calmette, who had worked with Louis Pasteur and his colleague the veterinarian Camille Guérin. After apparently successful tests, the vaccine was taken up in France in 1924, and several other European countries

THE LÜBECK DISASTER

In 1930, in the town of Lübeck in northern Germany, some 250 newly born infants were given the BCG vaccine against tuberculosis. Within weeks, over 60 had died, and about 80 were gravely ill. Some investigators believed that local doctors had been experimenting with their own version of the vaccine, which turned out to have fatal results. Despite two inquiries and a sensational trial, the truth was never established.

followed suit. In the English-speaking world, however, the vaccine was greeted with suspicion. This difference was most sharply emphasized in Canada, where the BCG was adopted by French-speaking Quebec in 1926, but not used in the English-speaking provinces until 20 years later.

Calmette and Guérin's vaccine consisted of live but weakened forms of *Mycobacterium tuberculosis*, the organism that causes tuberculosis. The bacterium usually infects people by being inhaled, and often lodges in the lungs. Here, it is engulfed by immune-system cells, and is normally destroyed. But in people whose defenses are weakened—for example through poor diet, overwork, crowding or stress—the immune-system cells engulf the bacteria, but fail to break them down. The bacteria can then reproduce, creating a lump called a

tubercle, which gives the disease its name.

The division over BCG concerned its safety. Calmette, in particular, energetically promoted his new vaccine, claiming spectacular results in France. Unvaccinated infants had a one-in-four chance of catching the disease if their mothers were infected, but in vaccinated infants the figure dropped to zero. These figures were so remarkable that Calmette's critics were unable to accept them. It gradually became clear, however, that the statistics were part of a broader picture. Calmette had monitored the progress of only a fraction of the vaccinated babies included in his report, so the true death rate could have been far higher than the zero level he had claimed. Eventually, he revised the figure to just under 2 percent, but by this time his statistics, and the methods used for compiling them, were under constant attack from abroad.

In 1928, this medical controversy reached the League of Nations,

TAKING SAMPLES Outside Emil von Behring's laboratories in 1903, visiting scientists watch as serum is drawn from a calf as part of research into tuberculosis.

which decided to adopt the vaccine for general use. Medical establishments in the United States, Britain and English-speaking Canada ignored the move. In 1930, in the German town of Lübeck, disastrous mistakes led to the death of more than 60 recently vaccinated children, providing potent

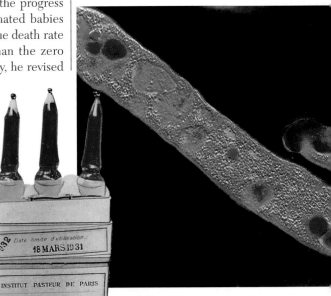

TARGETING A KILLER The BCG works by weakening the tuberculosis bacterium (above) to the point where it triggers immunity, without causing disease.

1908 BCG vaccine developed to treat tuberculosis

1918 Start of worldwide influenza epidemic

1923 Toxoid to treat diphtheria is produced

1931 Viruses are trapped; they are smaller than bacteria

1937 Widespread vaccination against yellow fever begins

ammunition to Calmette's increasing band of enemies. From then on, despite a growing list of favorable results from countries as diverse as Sweden, Poland and Spain, Calmette's critics were not to be swayed.

Eventually, in the 1950s, positive results became so overwhelming that the tide of opinion turned, although not before new drugs had arrived that offered ways to bring established cases of tuberculosis under control. Nearly half a century later, the disease is still far from conquered, but after the decades of argument and discord, Calmette and Guérin's vaccine continues to be used.

Diseases without a cause

Despite their differences, diphtheria, tetanus and tuberculosis have one thing in common: like many other infectious diseases, they are caused by organisms that are visible through a microscope. It was already clear by the beginning of the century, however, that this was not true of all diseases. Pasteur, for example, was unable to track down a bacterium that caused rabies, and no one managed to find bacteria that caused influenza, measles or the common cold. Pasteur's conclusion was simple and straightforward: he decided that the disease-causing organisms were merely too small to be seen.

Just how small soon became apparent. In the late 1890s, a Dutch bacteriologist, Martinus Beijerinck, investigated mosaic disease in tobacco plants and found that the infective agent could pass through filters with holes small enough to trap any known bacteria. He called the infective agent a "filterable virus," from *virus*, the Latin word for "a poison." In 1901, an American researcher, James Carroll, showed that yellow fever was caused by the same kind of agent, making one of the first connections between viruses and human disease.

This early work showed that, in all probability, a range of viral diseases might be treatable by vaccination. The difficulty was that, initially at least, viruses could not be investigated by growing them in dishes on a laboratory bench. Unlike bacteria, they turned out to depend on living cells, which meant

FATAL FLIES Spraying stagnant water with insecticide in Panama (right) in an attempt to control yellow fever and malaria, diseases found to be spread by mosquitoes (above).

that they could be grown only in animals—an expensive and laborious procedure.

In the early 1930s, Ernest Goodpasture, an American pathologist, came up with one answer to this problem. A fertilized hen's egg contains a complete living animal, packed inside a case that excludes most bacteria. The developing embryo inside has its own food reserves and immune system, so it can provide a perfect environment for viruses, and for forming antibodies against them. Then, in the 1940s, the invention of antibiotics cleared the way for researchers to use pieces of artificially cultured living tissue, which before had too often been attacked by bacteria, as a second source of virus-supporting cells. With these two virus "factories" on hand, the hunt for vaccines began in earnest.

One of the first products of this new field of research was a vaccine for a disease that spreads with the help of insects, instead of directly from person to person. Before 1900, there had been speculation that yellow fever was somehow linked to mosquitoes. Carlos Finlay, a doctor in Havana, was convinced of the link, and at the turn of the century an American medical team arrived in Cuba to try to establish if his mosquito theory was true. The United States had recently won Cuba from Spain, and yellow fever had bro-

ken out among the troops. Mortality from the disease was rising at an alarming rate.

The commander of the Army Yellow Fever Commission was military surgeon Major Walter Reed, who in 1900 worked with three medical colleagues—James Carroll, who discovered that yellow fever was indeed caused by a virus, Aristides Agramonte and Jesse Lazear. Reed spent most of his time in the United States, while his associates remained on the island. Agramonte, Lazear and Carroll used themselves as guinea pigs in their own research. All allowed themselves

to be bitten by mosquitoes that had already bitten infected people, and then awaited the results. Agramonte and Lazear were unaffected, but Carroll contracted the disease and nearly died.

The result was compelling, but did not yet constitute proof that the disease was transmitted by mosquitoes. After these initial experiments, tests were carried out on volunteer soldiers. These showed that mosquitoes carried the disease and that yellow fever could not be spread by dirty clothes or bedding, as had once been thought. Lazear later caught the disease through an accidental bite and died from its effects. Walter Reed, who did not take part in the tests, received widespread acclaim for the mosquito discovery, but himself died in 1902 from a ruptured appendix.

The Yellow Fever Commission's research

1950

1953 Salk vaccine for polio successfully tested in the United States

1969 First reported outbreak of Lassa fever

1976 Ebola and legionnaires' disease identified

1981 First reported cases of AIDS

1986 HIV virus recognized as cause of AIDS

1997 Hong Kong bird flu identified

2000

immediately opened up an environmental method for cutting infection. The mosquito concerned, called *Aëdes aegypti*, lays its eggs in shaded, stagnant pools. By systematically obliterating these breeding sites, either by draining them or by covering them with a film of oil, the disease's lifeline—its insect carrier—could effectively be cut. This approach was adopted around Havana, and it produced almost immediate results. It was also applied in Panama during the American bid to build the Panama Canal, a feat begun by the French in the 1880s but abandoned after thousands of workers died from yellow fever. When an epidemic spread in Panama in 1906, environmental control measures carried out by the Americans soon brought it to a halt.

The first experimental yellow fever vaccines were developed in the late 1920s, but became widespread only after 1937, when the South African microbiologist Max Theiler developed a safe but weakened strain of the virus. Since then, yellow fever vaccine has proved highly effective, giving

protection for a long period with very few side effects. However, since Reed carried out his experiments, medical researchers have discovered a fact unknown at the century's start: yellow fever viruses occur not only in humans and mosquitoes, but in monkeys as well. Despite widespread vaccination and mosquito control, there is therefore always a large reservoir of viruses able to trigger the disease.

Salk, Sabin and polio

Throughout the 20th century, effective vaccines were greeted with enthusiasm and gratitude. But those that occasionally caused harmful side effects—particularly in children—had a very different reception.

The century's greatest success was undoubtedly the vaccine against smallpox, which already existed, albeit in a crude form, when the century began. In 1977, it became the only vaccine that has led to the complete eradication of a major disease. By contrast, attempts to produce a vaccine against

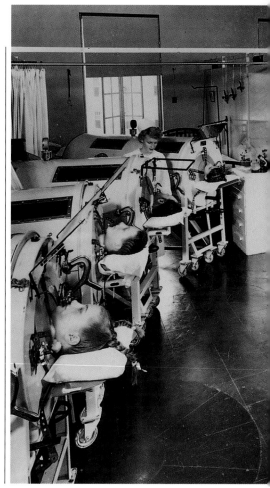

SMALL PRICE TO PAY A young boy receives a shot of Salk polio vaccine in the 1950s. Later vaccines were administered by mouth.

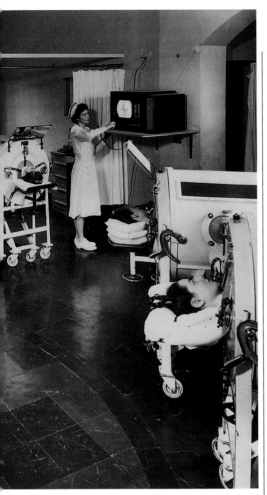

another viral menace—polio, or infantile paralysis as it was once called—ran into early difficulties, and only later went on to achieve success.

Polio largely affects children and young people, and was recognized as a viral disease in 1909. The first vaccines were not developed until the mid 1930s, at a time when the disease periodically reached epidemic proportions. Although most polio sufferers experienced few symptoms, some developed much more extensive complications, including paralysis of the muscles involved in breathing. In the 1930s, the only way that these patients could survive was with the help of an "iron lung," a machine that took over the work of the chest's paralyzed muscles.

Production of a vaccine proved difficult, because humans are the only natural hosts of the polio virus. However, researchers managed to grow the virus in monkeys, and vaccine trials were carried out in the United States in 1935. More than 15,000 children were vaccinated with a form of the virus that

had been chemically treated to make it inactive. Something went wrong with the vaccine, however, and the disease spread from the vaccinated children to some who had not been treated, and six of these children died. There was an immediate outcry. The test was halted, and the vaccine withdrawn.

Confidence in polio vaccines was severely tarnished, and the next development in the campaign against the disease did not take place until the late 1940s, when American scientists succeeded in growing the virus in cells specially cultured in a laboratory. It was from this source that Jonas Salk began experiments to create a form of the virus that, while inactive, would trigger immunity. Even after successful trials he was reluctant to propose its general use without further tests. But the prospect of overcoming such a calamitous childhood disease proved too much for the authorities overseeing his work, and in 1953 a nationwide vaccine trial

was carried out. This time, the results were unequivocal: infection rates plummeted, and there was not a single polio-related fatality in the children who had been treated.

Salk's original vaccine required booster doses in order to maintain immunity, and since the early 1960s it has been largely replaced by a different vaccine, devised by the Russian virologist Albert Sabin. Unlike the Salk vaccine, this is usually taken by mouth, and contains three living strains of the virus that have been weakened to prevent them causing the disease. In 1950, more than 5,000 new cases of polio were reported in the United States; by 1986, after three decades of vaccination, the figure dropped to three. A similar fall took place in many other countries.

Viruses in disguise

In marked contrast to polio, measles and smallpox, the struggle against many other

THE ARREST OF TYPHOID MARY

Some people can harbor disease-causing bacteria or viruses and pass them on without suffering from the disease themselves. In the 20th century, the most famous of these disease-carriers was an American woman named Mary Mallon, better known as "Typhoid Mary." Mallon worked as a cook in the New York area. In 1904, she triggered an epidemic of typhoid at Oyster Bay on Long Island, but disappeared when suspicion began to focus on her. In 1907, she reappeared in Manhattan, and again fled when she was dis-

covered. However, this time the New York medical authorities caught up with her, and she was committed to an isolation hospital on North Brother Island, off the Bronx. Mallon was allowed off the island in 1910, on condition that she did not handle food. However, she again worked as a cook, and caused more typhoid cases. In 1914, she was tracked down to a house in Westchester County, where she was living under an alias. She was sent back to North Brother Island, where she died in 1938.

ISLAND EXILE Mary Mallon, second from right, during her final years on North Brother Island.

THE CASE OF THE SNEEZING FERRET

DESPITE DECADES OF RESEARCH SPURRED BY SEVERAL EPIDEMICS AFFECTING MILLIONS OF PEOPLE, A RELIABLE ANTIDOTE TO THE FLU VIRUS REMAINS AS ELUSIVE AS EVER

In October 1918, just a month before the end of the First World War, the *Ottawa Journal* reported that "street cars rattle down Bank Street with windows wide open and plenty of room inside. Schools, vaudeville theaters, movie palaces are dark; pool halls and bowling alleys deserted." The center of the city seemed almost abandoned, not because of the war but because of an unprecedented epidemic of influenza.

The 1918 flu epidemic is estimated to have killed more than 20 million people— more than the war itself. The progress of the disease was rapid and particularly severe in young people. A doctor in Massachusetts watched helplessly as the victims were brought in. "Two hours after admission they have mahogany spots over the cheek bones, and a few hours later you can begin to see the cyanosis [blue discoloration] extended from the ears and spreading all over the face, until it is hard to distinguish the colored man from the white. It is only a matter of a few hours then until death comes, and it is simply a struggle for air until they suffocate."

ANTI-FLU SPRAY In 1920, bus and train interiors were sprayed in an attempt to halt the virus.

PROTECTIVE MEASURES Face masks (above) and gargling with antiseptic (right) probably did little to prevent the spread of flu.

At the time of the epidemic, very little was known about influenza, or the way it spread, but it was thought to be restricted to humans.

In the 1930s, a British researcher infected a ferret with the human flu virus. The reverse process, however, did not seem to work, apparently ruling out transmission from animals to humans. Then one of the researchers, Dr. Charles Stuart-Harris, happened to pick up a ferret, which sneezed in his face. He promptly developed the disease.

After the Second World War, the full complexity of the flu became apparent. Researchers discovered that different flu viruses attack a wide range of creatures, including pigs, horses, ducks and geese. The sudden emergence of new strains of the flu is thought to be caused by genetic recombination—chemical exchange between strains, creating different variants. Once a new strain has formed, migratory birds can then spread it around the world.

viral infectious diseases has been much less successful. For example, after decades of research, there is still no vaccine against the common cold, or one that can reliably prevent influenza. The reason for this, as immunologists discovered in the middle of the century, is that these viruses keep changing their chemical nature.

By the early 1930s, medical researchers had identified over three dozen viral diseases, but there were still few clues to indicate what viruses actually were. Some workers in the field suspected that viruses were individual molecules—collections of atoms so small that they could never be filtered out of liquid solutions. But in 1931, an English bacteriologist, William Elford, managed to trap viruses in membranes acting as extremely fine filters. From this, he deduced that viruses were particles—bigger than molecules, but far smaller than most bacteria.

Elford's breakthrough led to further discoveries about these enigmatic objects. They could be dried out, crystallized and stored on a laboratory shelf—even for years—without losing their power to cause disease, and they turned out to consist mainly of proteins, together with a small amount of nucleic acid. At the time, the significance of this discovery was not fully appreciated, because the function of nucleic acids was still unknown. In 1944, however, three American biochemists—Oswald Avery, Colin Macleod and Maclyn McCarty—showed that one nucleic acid, DNA, is the substance that controls living cells. Far from being inert packages of chemicals, as some had thought, viruses now looked suspiciously as if they might be alive.

Viruses were first seen in the mid-1940s, with the help of a recently invented instrument, the electron microscope. These early pictures, showing the tiny particles and their angular shadows, revealed that viruses are built with almost mathematical precision, with protein cases enclosing the nucleic acid within. Later pictures, taken in the 1960s, showed more detail still: some viruses turned out to be

DISCOVERING RICKETTSIAE

In 1908, while investigating Rocky Mountain spotted fever, the American pathologist Howard Ricketts discovered a new and remarkable group of bacteria. Named *rickettsiae* in his honor, they are not much bigger than the largest viruses. Unlike other bacteria, they are parasitic, and cannot reproduce outside living cells. Ricketts discovered that these unusual bacteria are the cause not only of Rocky Mountain spotted fever, but also of typhus. During his research he contracted typhus, and died from it in 1910.

enveloped in membranes, while many were studded with tiny chemical spikes. These spikes help a virus lock onto the cells it infects, but they also do something else: by subtly altering as time goes by, they present vaccines with a constantly changing target.

The history of influenza research shows how taxing this problem can be. In 1933, virologists succeeded in isolating the first chemically identifiable virus strain, which remained widespread until 1946. In 1947, a different strain appeared, with new kinds of chemical spikes. In 1957 this was supplanted by another strain, which caused a global epidemic of "Asiatic flu," and in 1968, a similar event occurred, with the arrival of yet another variant, labeled "Hong Kong flu." In 1997, a new strain called "chicken flu" was identified in Hong Kong. Vaccines have been developed against many individual strains, and some epidemics have been partly stemmed. But because the virus changes so quickly, and because many types are in circulation, flu prevention remains an impossible task.

New century, new diseases

Throughout the latter decades of the century, newspaper headlines have reported a num-

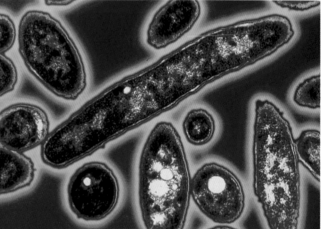

UNWELCOME VISITOR The Bellevue-Stratford Hotel in Philadelphia attracted unwanted fame due to the deadly presence of the legionnaires' disease bacterium (above).

ber of apparently new diseases that have struck without warning, sometimes with deadly results. One of the earliest was Q-fever, a mysterious flu-like illness first identified in Australia in the 1930s. Q-fever turned out to be caused by a rickettsia—an extremely small type of bacterium that lives parasitically inside the body's cells. Another of these "new" diseases made headlines in 1976, when 29 people died of a lung infection after attending a meeting of the American Legion in Philadelphia. Now known as legionnaires' disease, this infection was traced to a hitherto unknown bacterium that thrives in the water tanks of air-conditioning systems.

Although they were new to science, nei-

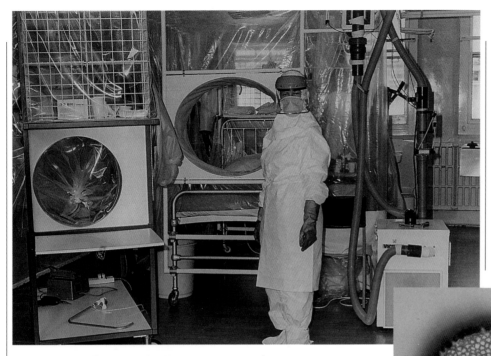

TOTAL ISOLATION This unit in a London hospital (above) was used to isolate victims of the Lassa fever virus (right). Food is passed to the patient via the small unit on the left.

ther of these diseases was actually new in the true sense of the word. The bacteria that cause them are likely to have existed for millennia, although in the 20th century changes in human habits and distribution gave them a better chance of staging an attack. The same is true of some viruses, but here the picture is more complex. Because viruses can evolve remarkably quickly, they are even better placed to exploit changes in the way we live.

Before the Second World War, when much less was known about viral disease, many rare forms of viral infection probably escaped detection. But by the 1960s, when a crop of "new" viral diseases broke out, they received detailed medical attention. One, called Marburg disease, affected a group of medical researchers in Germany, killing seven. The virus that caused it turned out to be found in the blood of monkeys, which were being used in research. Another, Lassa fever, which was first reported in 1969, broke out in West Africa, and subsequently spread to Europe and North America. It was traced to a virus found in forest rats, which had come into contact with people through

changing patterns in human settlement. Ebola disease, which was identified in 1976, also had its source in Africa and was linked to changes in human habits. Ironically, in this case the trigger was probably a new hospital, where the virus spread through infected needles.

These three viral diseases, and several others, have periodically flared up, only to flicker and die away. But in 1981, in Los Angeles, a viral disease emerged that did not follow the typical pattern. Instead of fading away, the disease—now called Acquired Immune Deficiency Syndrome, or AIDS—went on to become one of the most significant and widespread epidemics of the 20th century. Since AIDS was first recognized as a viral disease, immense efforts have been made to find a way to stop the virus in its tracks. But the task is not easy. Instead of attacking cells in vulnerable parts of the body, such as the throat or lungs, the virus goes to the heart of the body's defenses and attacks the immune system itself. This lays sufferers open to a host of "opportunistic" infections, which are triggered by viruses, bacteria and other microorganisms that, under normal circumstances, the body can keep in check. To make matters worse, the virus has a remarkable ability to change its chemical nature, excelling that of the flu virus many times over, which makes effective vaccines very difficult to produce.

As the AIDS epidemic nears the end of its second decade, the numbers of people infected with the virus are declining in the industrialized world, but they are rising fast elsewhere. Chemical treatments offer increasing hope to existing AIDS patients, but after all the successes with other infectious diseases, the prospects for prevention still seem remote.

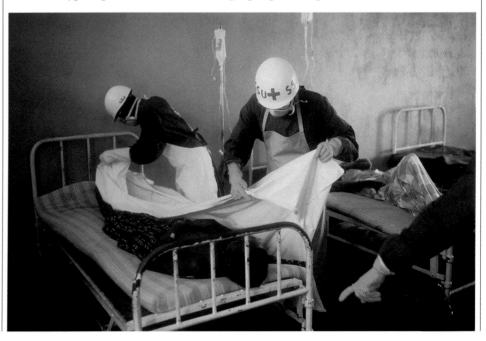

DEATH IN THE TROPICS In a clinic in central Zaire, patients dying of Ebola disease are tended by nurses in protective clothing.

THE BATTLE AGAINST AIDS

IN 1981, FIVE CASES OF A PREVIOUSLY RARE LUNG INFECTION IN THE UNITED STATES ATTRACTED ATTENTION AND PROVED TO BE THE TIP OF A TERRIFYING ICEBERG

The discovery of AIDS, or Acquired Immune Deficiency Syndrome, can be traced back to a strange coincidence that came to light in June 1981, when reports to the Centers for Disease Control (CDC) in Atlanta highlighted five cases of a previously rare lung infection caused by a microorganism called *Pneumocystis carinii*. Until 1970, this microbe was not even known to cause disease, so these cases of *Pneumocystis pneumonia*—based in the Los Angeles region—quickly attracted attention. More remarkable still was an immediate link between the five sufferers: all were men, and all were known to be homosexual.

Further cases were reported, bringing the totals up to dozens and then hundreds; patients included women, hemophiliacs and drug-users. Many patients were suffering from a range of other infections in addition to *Pneumocystis pneumonia*, again often caused by microorganisms that are normally innocuous. It became apparent that something had disabled the victims' immune systems, leaving them vulnerable to microbes that the body normally keeps in check.

Within months of the first reported cases of AIDS, the search was

WORLDWIDE THREAT Within a decade of being identified, AIDS had spread across the globe. This AIDS awareness poster is from Laos.

on to track down its cause. In 1983, Robert Gallo at the National Cancer Institute cloned the genetic fingerprint of a virus called HTLV-III, which he thought was involved. At the same time in Paris, Luc Montaignier announced that AIDS was caused by a virus he called LAV. After considerable wariness of each other's work, and rival claims for making the discovery first, the two men agreed that their viruses were identical. In 1986, a new name was devised: human immune deficiency virus, or HIV, officially became the cause of AIDS.

Research showed that HIV is a paradoxically fragile entity. Unlike many other viruses, it is easily destroyed outside the body, and cannot survive even the weakest of disinfectants. However, once inside the body, its grip is tenacious. It invades helper T cells in the immune system, and can remain dormant inside them for many years. These cells are vital to the body's defenses, and it is this feature of the infection that leaves the body open to further attack. By 1986, more than 20,000 cases of AIDS had been identified across the world. By 1996, the figure had risen to many millions, with the fastest increases occurring in developing countries.

Since the virus was discovered, a huge amount of work has gone into creating a vaccine that might arrest its spread. As the century closes, that work has yet to be successful, mainly because of the immense obstacle of ensuring a vaccine's safety. In 1994, a group of British AIDS researchers discovered that weakened viruses, developed for a similar disease that affects monkeys, could spontaneously "repair themselves," and become highly virulent once more. Until this can be ruled out for HIV, there is little likelihood of a safe live vaccine becoming available.

HELPING HAND Diana, Princess of Wales, meets a patient in a Canadian hospice in 1991. Diana's work with AIDS sufferers helped foster a broader understanding of the disease around the world.

THE CHEMICAL ARMORY

PATIENCE, PERSEVERANCE AND LUCK HAVE YIELDED AN ARSENAL OF POTENT CHEMICAL WEAPONS IN THE FIGHT AGAINST DISEASE

At the beginning of the 20th century, medicine production could be a highly profitable business. Patent medicines claimed to cure all kinds of diseases and disorders, and some made a fortune for their inventors. But behind the lavish advertising lay a startling fact: the number of specific drugs known to have a beneficial effect could be counted on the fingers of two hands. In a world of mixtures, tinctures and

WINNING TEAM The work done by Paul Ehrlich and Sahachiro Hata produced a breakthrough in the search for a synthetic drug effective against syphilis.

elixirs, which were often derived from a range of different ingredients, chemically pure drugs were very rare. The science of immunology was making rapid advances, but progress in pharmacology—the science of drugs—was slow.

To the average doctor in 1900, the contrast between these two branches of medicine must have seemed profound. Immunologists had achieved some spectacular successes in combating diseases such as diphtheria, but drug therapy still had mixed results. More than this, there was a difference in focus between the two kinds of treatment. A vaccine or an antitoxin seemed to find its target with uncanny accuracy, like a well-aimed bullet. By comparison, the action of drugs often seemed vague and imprecise.

One leading German immunologist, Paul Ehrlich, found these differences puzzling. By 1900, Ehrlich was already a key figure in immunology, but his research gradually led him away from the immune system itself, with its complicated biological chemicals, to the very different world of chemicals that were simple enough to make in the laboratory. He reasoned that if it were possible to synthesize a chemical equivalent of the immune system's "magic bullets," it might be possible to kill bacteria but leave the body unharmed.

The challenge lay in knowing where to start, because without some kind of lead, the search was almost bound to end in failure. It was here that Ehrlich's past experience paid rich dividends: instead of investigating chemicals related to existing drugs, he began his quest with something completely different.

Compound 606

During his research into immunology, Ehrlich had become very familiar with the use of chemical stains and dyes, which make cells easier to see under a microscope. Initially, these came from natural sources, but during the second part of the 19th century, German chemists made great strides in creating new dyes by synthetic means. In microscopy,

an ideal dye is one that is highly selective, staining particular kinds of cells or structures and leaving others unaffected. Ehrlich realized that the principles that applied to dyes almost certainly applied to drugs as well. He believed that just like a dye, an effective drug must become "fixed" to its target, with the target differing according to the

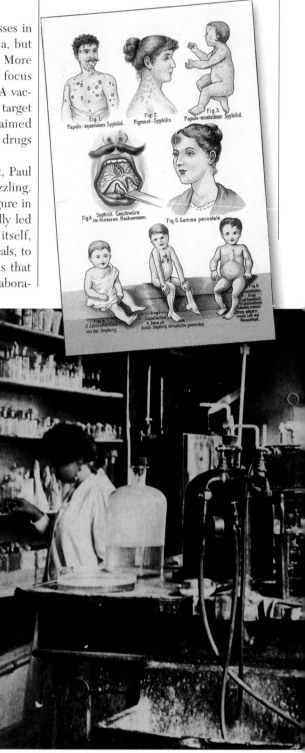

1904 Trypan red cures mice infected with sleeping sickness

1909 Compound 606 kills syphilis bacteria without harming humans

1922 Insulin first injected into a human suffering from diabetes

1929 *Penicillium notatum* identified by Fleming

1932 Prontosil found to be effective against streptococcal bacteria in mice

1941 The word "antibiotic" is coined

1944 Cortisone is first synthesized

drug concerned. By using dyes as a spring-board, he thought it might be possible to uncover a substance that delivered a fatal blow to the organisms that cause disease.

Ehrlich's first trials on human subjects involved a stain called methylene blue, which was known to stain the parasites that cause malaria. He used it on two people suffering from the disease, and although neither were cured, both showed signs of improvement. Ehrlich broadened his research to include a wide range of synthetic dyes and their derivatives. The work was arduous and also repetitive. When a promising substance was identified, it would often turn out to be too toxic for medical use, and the only way to get around this problem was to test dozens of derivatives, in the hope that one of them would turn out to be effective, and safe.

In 1904, Ehrlich reported that a dye called trypan red cured mice infected with

SECRET PLAGUE A German leaflet from the early 1900s (left) illustrates the symptoms of syphilis. Work in Ehrlich's laboratory (below) slashed the incidence of the disease.

THE CONQUEST OF SYPHILIS

At the time when Paul Ehrlich began his work on synthetic drugs, syphilis was a widespread and highly dangerous disease. Unlike diseases associated with poverty, it affected people from all walks of life, but its mode of transmission—through sexual contact—meant that it was rarely discussed. The origins of syphilis are mysterious. In Europe, the first reliable records of the disease date back to the 1490s, and there is some evidence that it may have been brought to Europe by Spanish soldiers who accompanied Christopher Columbus on his second journey to the Americas.

The cause of syphilis—a corkscrew-shaped bacterium called *Treponema pallidum*—was discovered by Fritz Schaudinn and Eric Hoffmann in 1905. In 1906, the immunologist August von Wasserman devised a test for the disease, which revealed it even in people with no apparent symptoms. This proved to be extremely useful, because unlike many infectious diseases, syphilis has a long "latent stage," during which all symptoms seem to vanish, only to reappear—sometimes years later—in a much more damaging form. As the bacterium spreads, it can eventually cause brain damage and a condition once called "general paralysis of the insane."

Ehrlich's treatment for syphilis—Salvarsan, or compound 606—was initially tested on spirillosis, a related disease that affects chickens. Trials in 1909 showed that it was lethal to spirillosis bacteria, and after tests on rabbits it was used on humans. The tests showed that it was highly effective at curing the early stages of the disease, but less successful once the disease had reached its final or tertiary stage.

Today, tertiary syphilis is rarely seen, but the disease still maintains its grim reputation. In 1997, a group of African-American sufferers received a public apology from President Bill Clinton, after the revelation that they had been left untreated for decades as part of a secret clinical test. Today, the disease infects less than 30 Americans in every 100,000 per year, and with the prompt use of antibiotics can quickly be cured.

MYSTERY UNMASKED Fritz Schaudinn correctly identified the spiral bacterium *Treponema pallidum* (above) as the cause of syphilis.

sleeping sickness parasites. In 1908, he discovered that a different compound—number 418 in a series of tests—had some effect on human syphilis, although it produced dangerous side effects. Then, in 1909, Sahachiro Hata, Ehrlich's assistant, tested a compound numbered 606, which had some characteristics of trypan red, in a formulation that also contained arsenic. It proved to be deadly to syphilis bacteria without harming the body. Ehrlich's hopes were realized: drug treatment, or chemotherapy as he called it, promised to stop one of the world's most devastating diseases in its tracks.

In April 1910, Ehrlich announced the results of his work. They caused medical pandemonium. Ehrlich's laboratory in Frankfurt was besieged by doctors begging for supplies of the new drug, and huge volumes of mail brought entreaties from overseas. Despite his initial clinical trials, Ehrlich was anxious about the drug's possible side effects, and insisted that it would not become generally available until further tests had been carried out. But the pressure for the new drug proved irresistible. Eventually, Ehrlich gave way, and production of the drug—now named Salvarsan—began on a commercial scale.

Ehrlich's triumph turned into something of a personal calamity. There was confusion about how the drug should be given—Ehrlich initially advised injections into muscles, before changing his advice to veins—and some doctors, lacking the skills required, bungled the procedure. The course of treatment was lengthy, and it needed to be tailored to each patient's condition, increasing the possibility of miscalculation. Although Salvarsan was far safer than the only medication that existed previously—highly toxic mercury—a number of patients died. Ehrlich devised a safer formulation, Neosalvarsan, which was released in 1912, but his troubles did not end there. His detractors accused him of profiteering, and also condemned him for encouraging licentious behavior by making it easier to cure a sexually transmitted disease.

The wilderness years

After Ehrlich's pioneering work, the stage looked set for a wave of further discoveries in pharmacology. But reality failed to match expectations. For the following two decades, remarkably little progress was made, apart from the discovery of drugs such as pamaquine naphthoate, which was used in the continuing struggle against malaria. The list of

---

THE NAMING OF ANTIBIOTICS

The term "antibiotic" was coined in 1941 by the Russian-American microbiologist Selman Waksman, who discovered streptomycin. An antibiotic is any substance produced in small amounts by one microorganism that inhibits the growth of another. Since Waksman's time, many of the antibiotics that were originally discovered in nature have been replaced by versions made synthetically.

---

effective drugs remained surprisingly short. A mood of disappointment set in, which was reflected in a report published in America in the early 1930s. Discussing the role of drugs in fighting tuberculosis, it concluded that Ehrlich's "magic bullets" had turned out to be more of a dream than a reality.

While further advances seemed elusive, these years of disappointment and stagnation did produce considerable success in a quite different area of chemical therapy. Early in the century, physiologists in Europe and America had started to uncover substances that act like chemical messengers, regulating the way the body works. They found that, despite their differences, these chemicals had many features in common: all were released directly into the blood by specific

The birthplace of Insulin.

glands, and all had very clear-cut effects. All they lacked was a name.

Today, that name—hormone—has become part of everyday language, but in 1905, when the word was first coined, the whole idea of chemicals controlling the body was new and unfamiliar. However, numerous experiments on animals soon showed that hormone deficiencies could cause diseases such as diabetes. From this starting point, it was a short step to the idea that hormone treatment might be able to effect a cure.

In the summer of 1921, two Canadian researchers at the University of Toronto, Frederick Banting and Charles Best, began work on a task that had defeated those working in endocrinology, as the new science of hormones and hormone-producing glands was called. Their particular interest was the pancreas, an organ that makes juices that play a part in digestion. Earlier researchers had discovered that the pancreas actually has a second role: as well as being involved in digestion, it also releases a hormone that controls the level of glucose in the blood. This hormone is produced by specialized clusters of cells called islets, and the Latin for an island—*insula*—gave the hormone its name.

That insulin existed, there was little doubt. But until Banting and Best began their collaboration, no one had managed to extract insulin from the islets that make it. It seemed to the two Canadians that the pan-

CRADLE OF DISCOVERY Frederick Banting (left) and Charles Best (right) pictured with the research institute at the University of Toronto where they developed insulin.

creas's dual role might lie at the heart of the problem, with the digestive juices breaking down the insulin before it could be removed. To test this theory, they operated on a dog, and tied off its pancreas from the rest of its digestive system. The part of the pancreas that made digestive juices shrivelled up, leaving the islet cells intact.

The decisive part of the experiment now began. The two researchers made an extract from the dog's islet cells, and injected it into another dog that was in the final stages of diabetes. That animal, close to death, immediately revived. On January 11, 1922, having first tested the extract on themselves, they went on to inject it into a 14-year-old boy who was at the threshold of death.

DIABETES TREATMENT Iletin insulin, made in 1923, and standardized insulin (left) produced in 1927.

Within days, to the astonishment and delight of his doctors, the boy was back on his feet.

Like Salvarsan before it, insulin had an almost instant impact in the medical world, and the demand for the hormone immediately outstripped supply. Production quickly moved from the university to two commercial laboratories, and from there to other sites throughout the world. However, unlike Salvarsan, insulin turned out to be a highly complex substance, and it could not—and still cannot—be made purely by chemical means. At first it was collected from animal tissue, but in the 1970s, the genes that control insulin production were successfully transferred to bacteria. Today, almost all insulin is produced by bacteria that are specially grown to ensure a steady and consistent supply.

Banting and Best's breakthrough transformed the lives of millions of diabetics. It also opened a way to the treatment of some other hormone-related disorders, and several decades later led to the development of the "pill," or oral contraceptive. But as has often happened in medicine, credit for the achievement was unevenly acknowledged. Banting was awarded the Nobel prize jointly with John J. Macleod, the professor who had authorized his work. Controversially, Best received nothing, and neither did another key figure, the biochemist James B. Collip, who helped to purify insulin extract in 1923. Banting publicly shared his part of the prize with Best, while Macleod shared his with Collip.

The mysterious birth of sulpha drugs

About two decades after Ehrlich's discovery of Salvarsan, one of his compatriots, Gerhard Domagk, made the second of the century's major advances in the chemical struggle against disease. But while Ehrlich immediately publicized his work, Domagk acted very differently. For reasons that are still obscure, he kept news of a great discovery secret for three whole years.

Domagk, a biochemist, was an employee of I.G. Farbenindustrie, a

PORTABLE PUMP Powered by battery, this infusion pump delivers insulin at a constant rate, imitating normal production of insulin in the body.

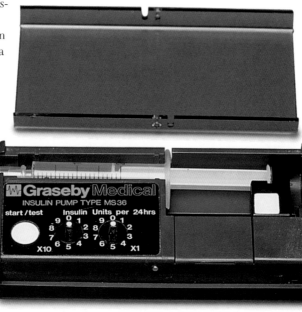

THE THALIDOMIDE DISASTER

During the early years of the century, there were few legal restrictions on the sale of medicinal drugs. The first significant piece of legislation was passed in the United States in 1938, after a scandal involving sulphanilamide, one of the new family of sulpha drugs used to fight bacterial infections. In the summer of 1937, the S.E. Massengill Company of Tennessee marketed a liquid form of the drug mixed with the solvent diethylene glycol, which is sometimes used as antifreeze. Within weeks, reports arrived of deaths connected with the mixture, and warnings were broadcast throughout the state. However, hundreds of people took the drug, of whom 107 died. In response, Congress passed a bill requiring every new or modified drug to be tested by the Food and Drug Administration before being licensed.

In the early 1960s, the FDA spared the nation from a drug disaster that enveloped a number of other countries, particularly in western Europe. Beginning in 1960, doctors began to report an unusually high number of birth deformities involving a previously rare condition called phocomelia, or shortened, seal-like limbs. After nearly two years of research, the cause of the deformities was identified as a drug called thalidomide, a sedative recommended for pregnant women who suffered from morning sickness. Women took it during the early months of pregnancy, when the fetus's heart, brain and limbs are formed—the time when it could exert a terrible effect.

More than 10,000 children were born with thalidomide-related deformities, and in the wake of this disaster the testing procedures for new drugs were tightened throughout the world.

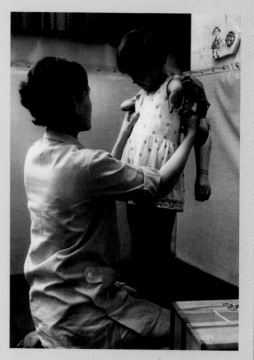

HELPING HAND A "thalidomide baby" is fitted with a prosthetic arm in a German clinic. Today, modern technology helps thalidomide victims.

giant German chemical company that specialized in producing industrial dyes. Following in Ehrlich's footsteps, Domagk carried out a series of tests on a new dye, called Prontosil, which had recently been developed for treating leather. In 1932, Domagk found that Prontosil was effective against streptococcal bacteria in mice, and it also seemed to be harmless to the mice themselves.

In humans, streptococcal bacteria are responsible for causing many kinds of infection, so the performance of Prontosil was particularly significant. This makes it all the harder to explain why Domagk was so unforthcoming about his work. Some have put it down to the exacting standards that he showed in his research, but there have been other interpretations for the delay. One is that Domagk knew that Prontosil could be converted into a more effective drug, called sulphonamide, which is formed when Prontosil breaks down inside the body. Sulphonamide had already been synthesized in 1908 by an Austrian chemist, so Domagk—aware that it was not new, and therefore could not be patented—might have been searching for a similar substance that would reap his employers rich rewards.

Whatever the truth behind this three-year interval, the outcome was decisive. When Domagk did eventually publish his work, the concept of "chemical bullets" gained a new lease on life. As other biochemists explored the potential of Domagk's discovery, the new drug, renamed sulphanilamide, became the

ACCIDENTAL HERO Alexander Fleming became a scientific idol after his discovery of penicillin.

first member of a rapidly growing pharmacological family of "sulpha drugs." In 1938, it was followed by sulphapyridine, which proved better at treating pneumonia, and other variations on the same chemical theme. The strengths and weaknesses of these compounds prompted research into a second, closely related, family of drugs, the sulphones, which had previously been dismissed because of their toxic side effects. One of these drugs proved to be the first effective treatment for leprosy, and another—isoniazid—became a standard treatment for tuberculosis from the 1950s.

Domagk succeeded in opening the door onto a wide range of chemical treatments; his initial discovery, sulphanilamide—used to combat disease-causing bacteria such as streptococci—saved huge numbers of lives. He was awarded the Nobel prize just after the outbreak of the Second World War, but after being arrested by the Gestapo, he declined the accolade. Eight years passed

POTENTIAL UNLEASHED Ernst Chain (right) and Howard Florey (far right) transformed penicillin from a laboratory curiosity into a life-saving medicine.

before he was able to accept it, during which time his breakthrough was eclipsed by a new class of drugs: antibiotics.

In 1939, the French-American microbiologist René-Jules Dubos was investigating life in the soil. He reasoned that as all organic matter eventually breaks down in soil, soil microbes must be able to dismantle or destroy cells of all kinds—including those of dangerous bacteria. During his research, Dubos discovered that a harmless soil bacterium, called *Bacillus brevis*, released a substance that could kill other bacteria nearby. When this substance was injected into mice, it killed disease-causing bacteria.

Dubos called the substance tyrothricin, and tried to interest other researchers in its promising properties. But the response was scant. At a meeting of American bacteriologists in late 1940, a proposed discussion on the subject was dropped because not enough participants wanted to attend. Dubos, however, was not the only one active in the field. As 1940 slipped into 1941, a team of researchers on the other side of the Atlantic were feverishly pursuing another source of antibacterial compounds. Their source was not another bacterium, but a mold.

An echo from the past

Fame, as Paul Ehrlich discovered, can be a fickle and volatile thing. In the history of 20th-century medicine, few people have experienced such heights of fame and then such sweeping reappraisal as Alexander

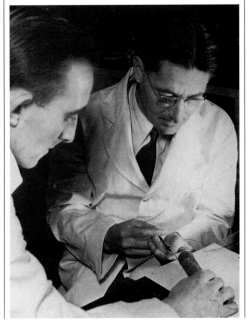

Fleming, the discoverer of penicillin. Long revered as the man who almost single-handedly created the world's most valuable drug, he is now seen more as one among equals—a man who made a crucial discovery, but failed to realize its full potential.

tantly, it had killed the bacteria living around it. He concluded that something seeping from the mold must be exerting a powerful antibiotic effect.

Fleming identified the mold as *Penicillium notatum*, a species that often grows on a wide range of organic substances. He carried out tests using extracts of the mold, and established that it was

MASS PRODUCTION Penicillin was produced on a large scale in France by 1946. The ampules below contain a standardized penicillin sample from 1945.

German-Jewish refugee, began a collaboration in Britain with Howard Florey, an Australian pathologist. In 1935, as part of a program of research into antibiotic action, they began a study into selected molds and bacteria. One of them was *Penicillium*. At first, the research was for purely academic interest, but as the Second World War approached, it developed a much more practical aim. Sulpha drugs had proved their usefulness, but had drawbacks. In time of war, more powerful antibacterial drugs would be immensely valuable.

Florey and Chain took up where Fleming had left off. They found that penicillin remained effective when diluted half a million times, and—contrary to Fleming's expectation—that it also remained active when injected into mice. On May 25, 1940, after a tiny amount of penicillin had been purified, Chain and his colleague Norman Heatley carried out a crucial experiment. They gave a potentially deadly dose of *Streptococcus* bacteria to eight mice, and then injected four of them with

NATURAL ALLY *Penicillium notatum* was the mold that yielded the world's first antibiotic.

Fleming was a bacteriologist at a London hospital. As part of his work, he cultured bacteria on dishes containing jelly-like agar, and then observed the bacterial colonies that were produced. In 1923 he isolated a substance called lysozyme from tears and mucus, and showed that it had some antibacterial action. However, with the exception of this discovery his career was uneventful, and but for a chance event in 1929, might have stayed that way. In that year, he picked up a culture dish that had been left out while he was on vacation, and noticed something odd: a mold had grown on the dish, but much more impor-

STEPPING UP PRODUCTION

During the Second World War, penicillin production increased in leaps and bounds. In early 1943, fewer than 1,000 people had been treated with the drug; by 1945, the number had risen to over 1 million. To generate this amount of the drug, researchers sought out variants of the *Penicillium* mold that produced unusually large amounts of penicillin. The most famous of these variants was found growing on a cantaloupe in a market in Peoria, Illinois.

effective against a variety of disease-causing bacteria. He also found that it was harmless to white blood cells. However, he was not able to isolate penicillin, and when he added the extract to blood in a test tube, it seemed to be inactivated, which suggested it would be useless in the human body. He published his discovery, and used penicillin to keep unwanted bacteria out of his culture dishes. And there, it seemed, penicillin would stay: a useful laboratory agent, but little more.

Penicillin's rebirth started six years later when Ernst Chain, a

THE GOLD RUSH YEARS

AFTER THE DISCOVERY OF PENICILLIN, MICROBIOLOGISTS SCOURED SOIL AND EVEN SEWAGE IN THEIR SEARCH FOR ORGANISMS THAT MIGHT PRODUCE USEFUL ANTIBIOTICS

When penicillin entered commercial production in the early 1940s, Ernst Chain, one of the two men responsible for this achievement, commented on the likelihood that other antibiotics might be discovered. "I could not believe that we would not find other substances of equal, perhaps even greater versatility than penicillin. It was just too much to expect that Fleming had stumbled on the only antibiotic of use to man—the odds of this happening were astronomical."

Chain's words were prophetic. Even before the end of the Second World War, microbiologists in the United States and unoccupied parts of Europe had joined in the medical equivalent of a gold rush, trying to find new sources of antibiotics. Guided by the idea of ecological competition— that microorganisms use chemical weapons in the struggle for living space—they searched habitats where this struggle was likely to be most intense.

The first and most fruitful habitat was soil. This was where Selman Waksman and his colleagues tracked down streptomycin in 1943, after he had been sent cultures of *Streptomyces griseus*, a fungus-like bacterium found growing on a chicken's throat. The bacterium proved to generate a powerful antibiotic, and Waksman and his team collected large numbers of soil samples from the poultry farm and used them to culture more examples of the microorganism. Waksman's work was exhaustive, and it threw up its fair share

DISEASE IN RETREAT Photographs from 1948 show the dramatic effect of streptomycin on a lung infection.

RUN TO GROUND Selman Waksman discovered that *Streptomyces griseoviridis*—one of the huge numbers of microorganisms in the soil—produces a valuable antibiotic.

of failures. In the early 1940s, he found several promising substances, such as actinomycin and clavicin. Although powerful against bacteria in the laboratory, they turned out to be too poisonous for use in humans.

Among the most unusual locations for an antibiotic discovery was one explored in 1945 by Giuseppe Brotzu, professor of bacteriology at a university on the Italian island of Sardinia. Instead of looking for antibiotic producers at random in soil, Brotzu focused his search on the one place where he felt new ones might be found—the sea around a sewage outfall pipe. During his search, he came across a fungus called *Cephalosporium acremonium*, which produces an antibiotic called cephalosporin. Since its discovery, cephalosporin—like penicillin—has spawned a number of derivatives with different antibacterial properties.

THE CLOSELY GUARDED SECRET

The development of penicillin, like the development of the atomic bomb, was a decisive factor in the history of the Second World War. For the research team working on penicillin, important moral issues arose about how widely the new drug was to be distributed.

In 1941, Howard Florey, one of the two men leading the effort to purify penicillin, heard that the Germans had been attempting to acquire samples of the drug. In a letter to a member of the British Medical Research Council, he wrote that it was "very undesirable that the Swiss and hence the Germans should get penicillin." Florey recommended that there should be a specific policy "not to issue cultures of *Penicillium notatum* to anyone with a possible enemy connection," adding that Alexander Fleming, the man who discovered the drug, should be informed of the decision. Florey's attitude was clear: penicillin should be used by the Allied side only.

Legally speaking, Florey was wrong, because Red Cross agreements signed in Geneva did not allow restrictions on the spread of therapeutic drugs. However, in the end, penicillin did remain the property of one side in the war. German scientists carried out their own research on the mold that produces penicillin, and their reports were even secretly sent to Japan by submarine, but they never managed to develop the technology to produce the drug.

penicillin. Heatley stayed on until the wee hours of the morning, and watched all the untreated mice die. All the treated animals survived. The following morning, the cautious Florey commented: "It looks like a miracle." For Allied soldiers in the Second World War, that indeed is what it was.

Breaks in the armor

The subsequent history of antibiotics was dominated first by the scramble to produce penicillin on a large scale, and then by the search for similar substances from other microorganisms. In 1944, this search yielded streptomycin, which was the first effective antibiotic to be used against tuberculosis. After the end of the Second World War, it generated a host of others, including the tetracyclines, described as "broad spectrum antibiotics" because they are effective against a wide range of bacteria.

Soon after penicillin entered commercial production, it seemed that chemotherapy—as Ehrlich had originally hoped—might

DRUG SEARCH **A physiologist tests a batch of chemicals on nerve tissue, to see if they have any useful effects.**

indeed overcome most kinds of infectious disease. But despite their undoubted power, the new antibiotics turned out to have major flaws, including the problem that some people had an allergic reaction to penicillin.

Another weakness, which researchers in the field clearly understood, was that the new drugs would work only against bacteria. They could not—and still cannot—prevent diseases caused by viruses, such as influenza, measles, polio, yellow fever, hepatitis and the common cold. This is because viruses behave in a different way than bacteria, and multiply by taking over the chemical machinery of living cells. Anything that kills viruses is likely to kill the body's cells as well.

After the Second World War, immunologists continued to make progress in the fight against some viruses, but chemical therapy lagged far behind. In 1957, a team of British bacteriologists, headed by Alick Isaacs, discovered that cells attacked by viruses produce substances called interferons, which help to prevent viruses multiplying. At the time, it seemed that interferons might be to viruses what antibiotics were to bacteria, but four decades later, that expectation is as yet unfulfilled. One reason is that unlike antibiotics, interferons work only for a short period, and they do not protect cells that are already infected. Also, they are highly specific: human cells are protected only by human interferons, which makes it difficult to produce interferons in large amounts.

Genetic engineering was used to make bacteria produce human interferons in the

early 1980s, but clinical trials have yet to match the glowing predictions from the recent past. More recent drugs, such as acyclovir, have overcome some of these problems, but viruses are still very difficult to treat by chemical means.

AWASH IN URINE

Although hormones often have profound effects, they are usually present in the body only in minute amounts. Once they have done their work, they are either broken down by their target cells or by the liver, or disposed of in urine. In 1931, the German chemist Adolf Butenandt processed more than 5,000 gallons of urine— enough to fill a small swimming pool—during his successful attempt to isolate the male sex hormone, testosterone. He produced about 0.002 ounces of hormone.

While work continued to devise elusive antiviral drugs, the postwar world of medicine began to discover shortcomings with the "wonder drugs" that they already had. In 1939, for example, almost all cases of bacterial meningitis and pneumonia responded well to sulpha drugs. By the early 1960s, the success rate was just one in two. The same was true of antibiotics: the more widely antibiotics were used, the more quickly bacteria seemed to outwit them.

During the last three decades, the problem of bacterial resistance has become acute, and has already generated hospital-based "superbugs," such as methicillin-resistant *Staphylococcus aureus* (MRSA), which are very difficult to control. Bacteria, not humans, are nature's greatest chemical experimenters. They constantly change, or mutate, as they reproduce, throwing up random variations in

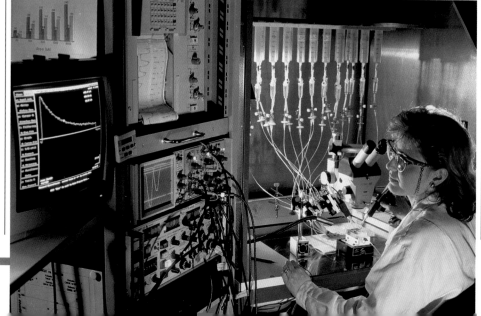

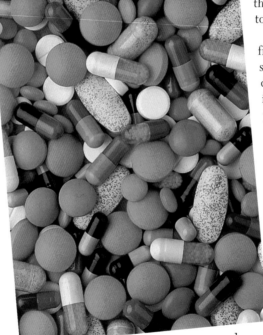

DRUGS IN ACTION A random streak (above) is all that remains of a bacterium dosed with an antibiotic. Today, dozens of drugs can produce carefully targeted destruction.

drugs. At the century's beginning, early pharmaceutical successes already included aspirin, which was actually first synthesized in the 1850s, although its therapeutic powers were not realized until the 1890s, and also paracetamol, which was first used as a drug in 1893.

Twentieth-century breakthroughs included cortisone, an anti-inflammatory first synthesized in 1944, and three important drugs that first appeared in the 1960s: propanolol, cimetidine, and allopurinol. Propanolol was the world's first "beta blocker," a substance designed to slow the heart and relieve the symptoms of angina, while cimetidine reduces the release of stomach acid, helping peptic ulcers to heal. Allopurinol prevents crystals of uric acid being laid down in the joints, thereby preventing the painful symptoms of gout.

These are just a handful of names from the immense number of drugs—some highly effective, others of doubtful value—generated by medical research during the last hundred years. During that time, the way that research is carried out has changed as much as drugs themselves. In the 1900s, pharmacology was mainly the province of university laboratories, and the processes involved were purely manual. Today, pharmacology is highly commercialized, and much of the work is automated, with computers taking the place of human technicians.

History also shows that while drug companies have expanded, and consumption of medicinal drugs has rocketed, on an official level at least, we are now much more cautious than we were. In 1910, within months of announcing its discovery, Paul Ehrlich was compelled to release Salvarsan for public use. In 1941, mainly due to the pressures of war, penicillin was made available within a similar period of time. But after several disasters triggered by drugs that were insufficiently tested, controls are now much tighter. Instead of waiting months, today's new drugs often must wait years before being licensed for use by the general public. Ehrlich's "magic bullets" now exist, but care is taken to ensure that none of them go astray.

their underlying chemistry. Most of these mutations are a handicap, so their owners soon die out, but just occasionally, one turns out to be useful. Because bacteria multiply so rapidly, a useful mutation is quick to spread. This is particularly true when a bacterium accidentally develops a mutation that protects it from an antibiotic. If it comes into contact with that particular drug, it will survive, while its relatives are likely to die. The outcome is that the resistant bacterium multiplies and spreads, making the antibiotic less and less effective.

Between the 1950s and early 1980s, indiscriminate use of antibiotics created many resistant strains of bacteria that remain problematic to the present day. Antibiotic use has now become much more precise and new antibiotics have been developed for resistant strains, but the cat-and-mouse game between antibiotics and their targets is likely to continue indefinitely.

Trials and errors

As Banting and Best demonstrated in the 1920s, when they isolated the hormone insulin, drug therapy consists of much more than a search for chemicals that will fight off the agents of disease. Many common diseases and disorders are caused not by infectious organisms, but by disturbances in the body's normal processes. On an increasing scale of severity, these problems range from transient nuisances, such as headaches, which afflict almost everyone at some time or another, to potentially life-threatening disorders such as hemophilia.

In the course of the century, the search for chemical treatments for disorders such as these has become one of the world's busiest fields of scientific endeavor. The result has been a flood of therapeutic

THE RISE OF MODERN SURGERY

SURGERY HAS CHANGED FROM A LAST RESORT TO ROUTINE TREATMENT THAT OFTEN HAS AN EXCELLENT CHANCE OF SUCCESS

Few professions in medicine have seen such changes as surgery. At one time, surgeons were looked on as little more than manual laborers, whose work was hurried, painful and bloody. However, by the end of the 19th century, they had shrugged off their former reputation. Thanks to the introduction of anesthesia and rigorously hygienic techniques, they had become the new medical elite, who dared to venture into parts of the body that their predecessors had left alone. In Vienna, for example, the pioneering professor of surgery Theodor Billroth had developed several new abdominal operations, working in a part of the body often left alone because of the dangers of infection, while in the United States, William Halstead—who introduced the use of rubber gloves into the operating room—had developed new surgical procedures for hernias, thyroid gland problems, and breast cancer.

In an era marked by recent successes, one German urologist remarked that the main work in his speciality had now been done. "For our successors," he predicted, "there remain only scanty gleanings." Ironically, during the 20th century, his field of surgery saw one of the most extraordinary developments of all: kidney transplants. However, before that kind of surgery could be contemplated, new skills and techniques had to be learned.

Gloves, masks and muscle relaxants

William Halstead's rubber gloves entered the surgical world by accident, after an operating-room nurse—who later became Halstead's wife—complained of an allergy to the disinfectants then in use. Halstead suggested the gloves, and discovered that they reduced the rates of accidental infection. By 1900, rubber gloves had become a standard item in many leading hospitals.

Other innovations were the result of observation and experience. The gauze face-mask, for example, was developed in Europe as a means of preventing infection through droplets of airborne mucus, while European surgeons also developed a range of new surgical clamps to reduce bleeding during operations to a minimum. Basic techniques were

CIRCULAR SAW This early 20th-century skull saw was powered by hand. A strong and steady grip was required to operate it.

also refined. Instead of cutting in any direction, as their predecessors often did, surgeons in the opening decades of the 20th century learned to cut parallel to the skin's natural creases and folds, which reduces the chances of a wound pulling open against its stitches. In a reversal of former practice, they also took their time. Until the

GETTING TO WORK A French surgeon operates on a patient in about 1900. Though clean, the room is not sterile.

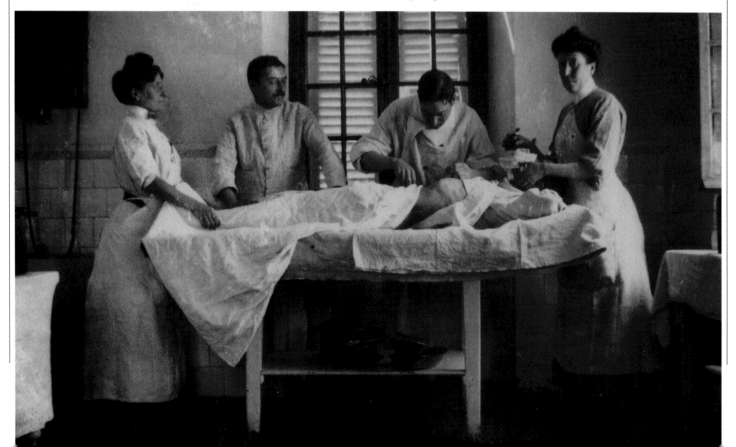

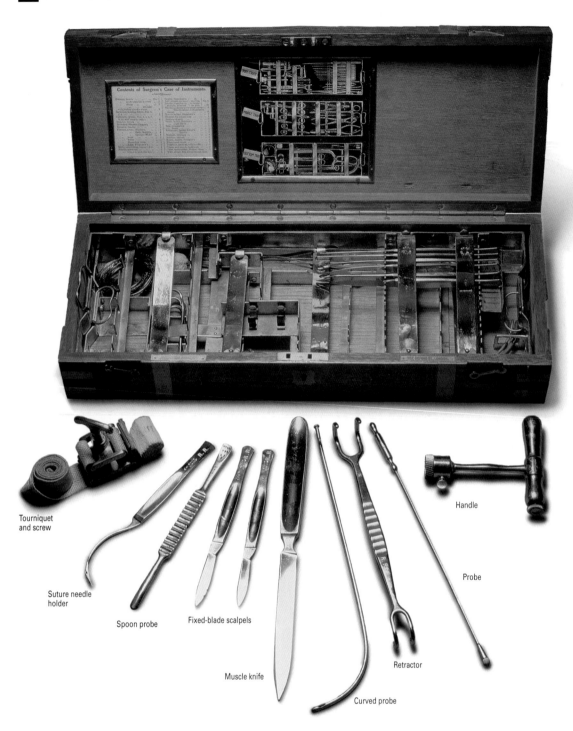

Contents of Surgeon's Case of Instruments.

Tourniquet and screw

Suture needle holder

Spoon probe

Fixed-blade scalpels

Muscle knife

Curved probe

Retractor

Probe

Handle

TOOLS OF THE TRADE This surgical instrument box was made in 1916. It was portable and intended for use in military hospitals during the First World War.

out on themselves. Although tubocurarine made some kinds of surgery easier, it also meant further changes in the operating room, because someone in the surgical team now had to operate the paralyzed patient's lungs. This extra work fell on the shoulders of anesthetists, making their work as important as that of the surgeon wielding the scalpel.

Alexis Carrell and vascular surgery

At the dawn of the 21st century, we take it for granted that surgeons can repair damaged blood vessels and sew back together ones that have been cut—a procedure that is now part of many major operations. Yet this area of work, called vascular surgery, is not as simple as it sounds. The fact that it can be done at all dates back to the pioneering work of one man, carried out just as the new century began.

In 1894, Alexis Carrel, then aged 21, was working as a surgeon in the French city of Lyon. On June 24, the city was visited by the French president, Marie François Carnot. During the visit, a man in the crowd stabbed the

poison wild animals—and their enemies. In the 19th century European scientists, amazed by its ability to cause instant paralysis, used curare in laboratory experiments. However, the drug remained no more than a chemical curiosity, and was not tested as an aid to surgery until 1912. The surgeon who used it, a German named Läwen, reported curare's value as a muscle relaxant, but little attention was paid to his work. Forty years then passed before two Canadian physicians, Harold Griffith and Enid Johnson, used it during an abdominal operation at a hospital in Montreal. They found that curare allowed the amount of anesthetic to be greatly

reduced, because the patient's muscles became completely slack and no longer resisted the knife.

Following this clinical test, the active ingredient in curare, called tubocurarine, gradually entered widespread use, although not before some alarming experiments that researchers carried

IT'S A KNOCK-OUT San Francisco's Director of Health demonstrates a new, purified form of nitrous oxide, introduced in the 1930s.

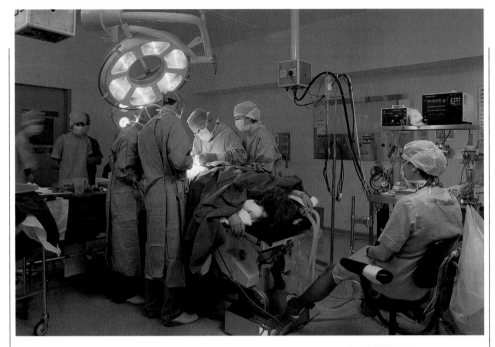

DRESSED FOR THE PART In a modern operating room, sterile gowns, masks and instruments reduce the chances of infection.

president, puncturing his liver and severing a major artery. Despite immediate medical help, Carnot died. The failure of medical science to save Carnot's life created a deep impression on Carrel, who decided to see if he could develop ways of repairing this kind of wound. For eight years he experimented with methods of joining severed blood vessels, and of closing them where they had been torn. It was a task fraught with difficulties. One problem was that of joining two severed ends when the vessel itself had collapsed. Carrel overcame this by a technique that involved a trio of stitches. When the stitches were pulled simultaneously, each severed end opened up into a triangle, making pairs of ends much easier to sew together.

Carrel encountered a more taxing problem concerning blood itself. White blood cells often cling to the inner surface of blood vessels as they seek out invading bacteria, but red blood cells do not. For them, the interior of blood vessels is a nonstick surface, ensuring smooth movement of the blood. But if a stitch intrudes into the interior of an artery or vein, it creates an obstacle in the otherwise slippery lining, and this can become the focus of a clot. Once a clot has formed, it may then become detached, and travel to the lungs or brain. The result often proves fatal.

In 1902, Carrel announced that he had

MEDICAL MAGICIAN A cartoon of Alexis Carrel pokes fun at his habit of carrying out experimental surgery on animals.

found a way of avoiding this much-feared complication. He rolled each severed end back on itself to create a short collar. The two collars could then be stitched together without any of the stitches penetrating the interior of the vessel itself. This technique was the first reliable form of blood-vessel repair, and it laid the foundations for procedures that are still in use today.

In 1904, Carrel moved to the United

States, where he continued his research. Spurred on by his ability to rejoin blood vessels, Carrel had started to use animals to experiment with organ transplants. All organs have their own network of blood vessels, but some—particularly the kidneys—are connected to the rest of the body by a very limited number of arteries and veins. For someone with Carrel's skills, it was relatively easy to cut through the vessels supplying a dog or cat's kidney, and to move the entire organ somewhere else, such as the neck, or even into another animal entirely. Carrel was not the first person to carry out these kind of experiments, but no one had pursued them with such rigor and zeal.

For many people—including some in the

SURGICAL FADS AND FASHIONS

Like other aspects of life, 20th-century surgery sometimes fell prey to changing fashions and misguided ideas. One of the most persistent was the idea that some parts of the body are vulnerable to "blockages" or infections, and are therefore better removed.

Early in the century, one exponent of this form of treatment was the British surgeon William Arbuthnot Lane. Lane believed that if wastes accumulated in the digestive system, they would eventually poison the body. He advocated surgery for anyone suffering from prolonged constipation, and removed substantial lengths of the small or large intestines. Today, this drastic procedure is carried out only in cases of intractable conditions such as cancer.

The century has also witnessed examples of mistakes by the medical profession as a whole. A prime instance concerns removal of the tonsils. From the 1920s until relatively recently, tonsils were considered potential nests of infection and surgeons removed them on the slightest pretext, despite the lack of any firm evidence that this did much good.

In his book *Limits to Medicine*, the sociologist Ivan Illich quotes a report from 1934 showing how far fashion can drive surgical intervention. A group of 1,000 schoolchildren were selected in New York; 61 percent had had their tonsils removed. The remaining 39 percent were examined by a group of doctors, who selected almost half to have their tonsils removed. The rest were seen by a different group of doctors, who selected more than 60 percent for surgery. The remaining children were examined by yet another group of doctors, and half were chosen for surgery, leaving just 65. The experiment was then halted because the supply of doctors had run out.

medical profession itself—Carrel's work was a ghoulish contravention of the laws of nature. However, for others, including the Nobel prize committee, it was valid and trail-blazing research. Carrel's investigations of grafts and transplants earned him a Nobel prize in 1914 and established two crucial points: it was technically feasible to transplant organs and tissues within a donor animal's own body, and in many cases, the animal in question would survive. However, if the organ

CORNEAL TRANSPLANTS

The most commonly transplanted part of the body is the cornea—the transparent, cup-shaped membrane that covers the central part of the eye, and which can become scratched due to scarring or disease. The cornea has no blood supply, and therefore no antibodies to cause rejection. The new cornea is sewn in place with a ring of nylon stitches. These have to stay in place for up to a year, because the cornea is slow to heal.

or tissue came from another animal, it would work only for a limited period before the animal died. Carrel had discovered the phenomenon of tissue rejection—something that was to prevent successful transplants in humans for many years to come.

Surgical specialization

While tissue rejection blocked progress in organ transplantation, surgeons were making strides in many other fields. One of these was neurosurgery. In the early 1900s, surgery of the nervous system was very much in its infancy. Surgery of the skull was an ancient practice, but surgery of the brain and nerves called for much greater knowledge and precision, and was a much more recent development. Some surgeons had succeeded in dealing with abscesses and tumors on the exterior of the brain, but in unskilled hands these operations were extremely hazardous: a slip of the scalpel could have irreparable consequences.

One man, the American surgeon Harvey Cushing, did much to promote neurosurgery as a specialized branch of medicine. Cushing became professor of surgery at Harvard University in 1912, and in the following 30 years he reduced the death rate during brain surgery to the unprecedentedly low level of just 5 percent. Cushing was a keen medical historian and passionate collector of books, but he also collected tumors that he had

removed from the brains of his patients. By studying and comparing these preserved specimens, he was able to classify the kinds of tumor that he had encountered and specify methods of surgical treatment.

In the period leading up to the Second World War, rapid progress was also being made in abdominal surgery, surgery of the eyes and ears, and in orthopedics, the branch of surgery dealing with bones and joints. Here, new methods of pinning broken bones internally were introduced, using plates and pins made of inert metals so that new bone would have a chance to form. No region of the body was beyond the reach of the surgeon's knife, including one of the most inaccessible of all—the thorax, or region enclosed by the chest.

As 19th-century surgeons discovered, operating on the chest is complicated by the problem of air pressure. The lungs do not have any muscles themselves; instead, they expand and contract as the chest changes shape. Around them is a narrow space called the pleural cavity, which is filled with a lubricating fluid. During normal breathing, the pressure in this space is less than the pressure of the atmosphere. This keeps the lungs stretched outward, so that air can enter them easily. However, if a surgeon deliberately or accidentally punctures this cavity during an operation, air rushes in and the lungs may collapse.

In the early 1900s, Ernst Sauerbruch, the founder of modern thoracic surgery, made an attempt at overcoming this apparently insuperable problem. With the help of a team of technicians in the German city of Breslau (now Wroclaw in Poland), he built a pressure-proof chamber that could accommodate a patient, an operating table and a complete surgical team. The patient's head protruded through an airtight seal to the outside, allowing air at normal pressure into the lungs. The pressure inside the chamber was then

reduced, so that it was significantly lower than that of the surrounding air. The surgeon—usually Sauerbruch himself—could open the patient's chest without any danger of lung collapse, because the air in the lungs was always under greater pressure than the air in the operating chamber.

The principle behind this "differential pressure technique" was faultless, and it allowed Sauerbruch to carry out surgical procedures, such as the removal of tumors around the lungs, which previously would have been highly hazardous. But working within the confines of a pressure-proof chamber was difficult, and Sauerbruch's invention did not catch on. By the end of the First World War it was replaced by a far simpler device: a tube with an inflatable cuff that could be passed down the windpipe and

SURGERY UNDER PRESSURE Designed in the 1960s, this high-pressure chamber increases the diffusion rate of oxygen through the lungs, giving surgeons extra time to operate on "blue babies."

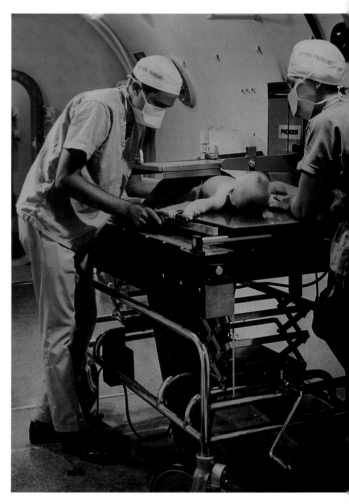

THE CASE OF "BABY FAE"

On October 26, 1984, a recently born infant became the world's youngest recipient of a heart transplant. Known only as "Baby Fae," she was given a baboon's heart in an operation at the Loma Linda Medical Center near Los Angeles. "Baby Fae" survived less than three weeks. In the furor that followed the operation, surgeon Leonard Bailey received numerous death threats from animal rights activists, and needed constant police protection. Baby Fae's operation effectively ended the use of primate donors in heart transplants.

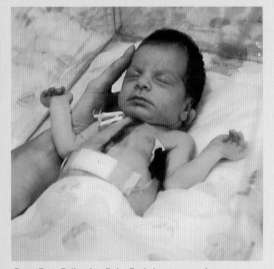

BABY FAE Following Baby Fae's heart transplant, hundreds of animal rights activists picketed the hospital where the operation was carried out.

then sealed in position. This could be used to deliver air under pressure, so that the lungs stayed inflated. More than this, it could also be used to deliver anesthetic gases—a feature that led to its widespread use, even in operations that did not involve opening the chest.

Journey into the heart

Once surgeons could safely open up the thorax, the way lay open to deal with problems affecting blood vessels around the heart, and even the heart itself. In 1921, the first successful operation was carried out to relieve a condition called constrictive pericarditis, in which the pericardium—a membrane around the heart—thickens and shrinks, making it harder for the heart to beat.

By the late 1930s, heart surgeons had grown more ambitious, and had moved on to correcting inherited abnormalities. In 1939, two surgeons in Boston carried out a new and life-saving operation on a young child, which involved tying off a blood vessel called the *ductus arteriosus*. This vessel is a kind of cir-

culatory short-circuit that prevents a fetus's blood flowing through its lungs before birth. Immediately after birth it normally closes, but in a tiny number of children—about 500 in every million—the duct remains open, leaving the child short of oxygen and exhausted by the slightest effort. Before the operation became feasible, most children with this condition had poor prospects, but with it, their chances were greatly improved.

In 1944, this operation was followed by one designed to help "blue babies"—children whose blood is discolored by a dangerous shortage of oxygen. Blue babies often suffer from a combination of four heart defects, known as the tetralogy of Fallot after the French professor who first recognized the condition. They include the so-called hole in the heart, when the partition between the two sides of the heart is incomplete. Surgeons could not correct these defects in the 1940s, but Helen Taussig, a Canadian pediatrician, suggested a way of replumbing the heart's blood vessels to improve blood flow to the lungs. Her ideas were put into practice on a 15-month-old baby, and saved his life.

Having gotten this far, surgeons were tempted to tackle problems that lay within the heart itself. The heart had to be kept beating, however, so there was no question of opening it up in any major way. Instead, surgeons made only small incisions, and worked on the heart's interior by touch rather than by eye. One of the most ingenious developments during this era of "closed heart" surgery was again made in Boston, by the American surgeon Dwight Harken. Harken operated on a patient with a narrowed heart valve, and treated the problem by scraping away at the valve with a tiny knife fitted on one of his fingertips. The operation was simple and effective, and is still sometimes used today.

Despite these advances, surgeons in the 1940s had reached a major barrier: they could not carry out any operation that involved opening up the heart, because if it was stopped for more than a few min-

utes, the patient would almost certainly die. Before operations like this could be attempted, a way had to be found of extending the time available to the surgeon. These methods did not appear until a new decade began.

Overcoming immunity

While surgeons continued to push back the frontiers of what was technically achievable on the operating table, research continued into the perplexing problem of tissue rejection. Carrel's experiments had shown that foreign tissue—wherever it came from—rarely survived being transplanted, and the first attempt to transplant a human organ, carried out by a Soviet surgeon in 1936, ended in failure. A female patient was given a kidney from a man who had recently died, but the kidney did not work, and the recipient herself died soon afterward. At this stage, the causes of tissue rejection were still unknown, and despite further attempts at transplanting human kidneys, the entire subject was seen as dangerous and medically unjustifiable.

During the Second World War, Peter Medawar, a British research scientist, started to throw some light on the process that

THE BODY AT WAR Peter Medawar was one of the first people to analyze tissue rejection—a problem that dogged early attempts at transplant surgery.

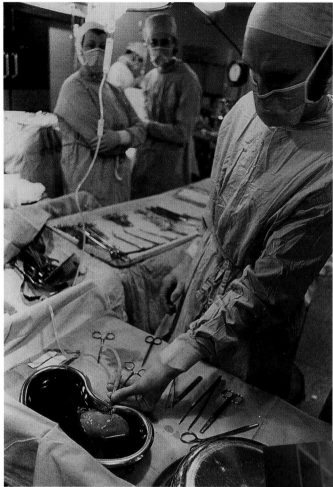

GIFT OF LIFE A human kidney is prepared for implantation. Tissue typing (below) minimizes the chances of rejection.

can change with time. In another series of experiments, he discovered a period of "immunological tolerance" in very early life, during which animals—and by extension humans as well—can accept foreign tissue that they would normally reject.

Despite their brilliance, Medawar's experiments did not have any direct application in the operating room, but they did suggest several ways forward in the fight against rejection. The immune system functions through white blood cells, so finding some way to disable these cells became a

made transplanted organs fail. Medawar was involved in work on skin grafts, which in wartime were often needed to help the victims of serious burns. The usual practice was to slice away pieces of the patient's unburned skin, and move them to areas where most of the skin had been lost, where with luck they would start to grow. However, in the severest cases, the technique would not work because there was simply not enough skin to spare.

While he was working on one patient, Medawar made two grafts—one from the patient herself, and another from a volunteer donor. As expected, the patient's own skin took, but the donated piece failed. Medawar then repeated the process, trying to graft a second piece of skin from the same donor. To his surprise, instead of gradually failing, the donated graft immediately began to die. The only possible conclusion was that the patient's immune system had learned to recognize the "foreign" tissue, just as it can learn to recognize invading viruses or bacteria.

Medawar's discovery had a profound impact, because it showed that far from being fixed, tissue rejection is something that

priority in transplant research. The most drastic method, first used in France in kidney transplants in 1958, involved bombarding the recipient's body with X-rays so that their white cells were completely destroyed. Replacement white cells from donors were then injected into the patient in the hope that they would not attack the transplant. However, with a few exceptions, the results were poor. Whenever rejection was successfully prevented, the recipients often died of infections that their new immune system could not overcome.

A second method of attack, first tested by several researchers in the early 1950s,

involved targeting white blood cells with drugs. At first, this form of treatment—called immunosuppression—had mixed results, because many immunosuppressant drugs turned out to have harmful side effects. But by the early 1960s, this form of therapy had begun to prove itself. Tissue

SHARED CIRCULATION

In the 1950s surgeons tried out a remarkable way of shutting down the heart during an operation—by circulating the patient's blood through someone else's body. Several successful heart operations were carried out on children in this way, using a parent to keep them alive. Critics described the technique as the only kind of operation that had a potential for 200 percent mortality.

rejection was gradually being overcome, and transplants had moved from being a reckless gamble to a rational form of treatment.

Since the first successful kidney transplant, carried out by surgeons in Boston in 1954, organ transplantation had been shown to work effectively, especially with new techniques of tissue typing, which help to match the tissues of donor and recipient. In 1963, American Thomas Starzl attempted the first liver transplant, but because the liver has a complicated blood supply, success did not come until 1967. In fact, Starzl's great achievement was soon eclipsed by another and far more emotive event in the history of 20th-century surgery: the transplantation of a human heart.

The surgeon as hero

By January 1964, the era of open-heart surgery was well under way. Thanks to the heart-lung machine—the invention of American surgeon John H. Gibbon Jr., which first appeared in 1952 and allowed a patient's blood to be oxygenated mechanically—surgeons could now stop the heart for an hour or more, allowing time for complex operations. Hundreds of patients had been successfully fitted with artificial heart valves to replace ones that were diseased, and the first attempts were under way at bypassing blocked coronary arteries, using pieces of vein taken from elsewhere in the body.

In this year, James Hardy, an American heart surgeon, became the first person to carry out a heart transplant on a human patient. The recipient was a 68-year-old man

CHRISTIAAN BARNARD: THE MAN WITH THE GOLDEN HANDS

For Dr. Christiaan Barnard, the world's first successful heart transplant proved to be a launch vehicle toward the kind of fame normally reserved for movie stars. Before the operation in December 1967, Barnard was a respected cardiac surgeon, but only one of a number worldwide who had contemplated carrying out the ultimate organ transplant. Outside his native South Africa, he was largely unknown. But once the operation had taken place, Barnard's life changed forever.

Despite the death of the recipient after just 18 days, Barnard became an instant hero. The handsome 45-year-old surgeon was soon a member of the international jet set, appearing on television shows in countless different countries, and meeting heads of state. The "Man with the Golden Hands" developed a playboy image, in marked contrast to his lifestyle in earlier days.

On January 2, 1968, Barnard carried out his second successful heart transplant. Unlike the previous recipient, Louis Washkansky, Philip Blaiberg made a strong recovery and was eventually allowed home. He survived for 18 months before symptoms of tissue rejection finally set in. Blaiberg died in August 1969, but despite the disappointment surrounding his death, his zest for life during his 18-month reprieve did much to boost the image of heart transplantation, convincing medical authorities that despite its immense cost, it was a worthwhile treatment.

Barnard's fame inevitably attracted a certain amount of jealousy in the world of surgery, particularly as he had snatched first place in the face of close competition from surgical teams in the United States. However, Barnard's professional reputation was not based simply on determination and good luck. Before the days of modern immuno-suppressant drugs, his attention to patient aftercare ensured that several of his patients managed to survive for more than a decade after their operations.

PLAYBOY SURGEON Christiaan Barnard had huge numbers of admirers. He helped to boost South Africa's image at a time when it was becoming tarnished by apartheid.

holds—is not a particularly complex organ. All that was missing was the fateful coincidence between a competent surgeon, a needy recipient and a suitable donor.

That coincidence occurred on December 2, 1967, in Cape Town, South Africa. While 53-year-old Louis Washkansky lay dying of heart failure in Groote Schuur Hospital, his wife drove past the site of an accident in which two pedestrians had been hit by a truck. One died outright; the other—a young woman—was so seriously injured that she had no hope of recovery. She was to become the donor that Louis Washkansky so desperately needed. Although her heart was still beating, a doctor at Groote Schuur certified that she was clinically dead. Her heart was transplanted by Dr. Christiaan Barnard, and within hours, news of the operation had flashed around the world.

Louis Washkansky survived for only 18 days before dying of a lung infection, but his transplant operation was like a shot from a

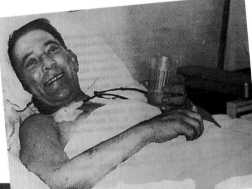

A WORLDWIDE FIRST Louis Washkansky looks cheerful after his heart-transplant operation. For several years, the surgical team at Groote Schuur Hospital (bottom) led the world in transplant techniques.

who had suffered a major heart attack, and Hardy's original intention was to transplant a heart from a donor who was on a life-support machine and who seemed to be at the point of death. But when everything was ready, Hardy suddenly found himself confronted by an appalling dilemma: the donor's heart was still beating, and the donor was therefore legally still alive. With a human heart unavailable, the only other source of a heart was a laboratory chimpanzee. He decided to go ahead.

In medical terms, the operation was not a success. Although the chimpanzee heart worked well after the operation, it was too small to pump blood around a human body. Hardy fitted pacemakers to boost the heart's activity, but the man died soon afterward. In public relations terms as well, the operation was a disaster. Hardy found himself at the center of an ethical storm, and few surgeons

tried to follow in his footsteps. However, the technical feasibility of heart transplantation no longer seemed in doubt, because the heart—despite the significance that it

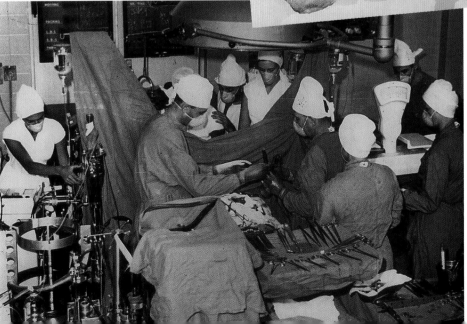

SELF-SEALING SCISSORS
These 1990s surgical scissors use an electrical current to seal off blood vessels as they cut through them.

starter's gun. On December 6, 1967—just four days later—the world's second human heart transplant was carried out by Adrian Kantrowitz on a child in New York. Exactly a month after that, Norman Shumway operated on a 57-year-old man, again in the United States. In both cases, as with Washkansky, the results were not good: the child died within hours; the man survived just 15 days.

By normal standards of medical appraisal, the outcome of these three operations should have given pause for thought. But critical judgment was suspended, and heart-transplant surgery developed its own unstoppable momentum. Cardiologists all over the world wanted their share of the fame it brought, and in 1968, more than 100 transplants were carried out. However, the problems of rejection and infection had not gone away and few recipients survived for long.

In the years that followed, the pendulum of public opinion swung from admiration to censure. After one failure, a medical correspondent wrote in the London *Times* that there could be no justification for an operation whose results were "more or less equivalent to a death sentence."

With the development of new immunosuppressive drugs, such as cyclosporin, in the early 1980s, the prospects for transplant patients of all

MICROSURGERY Eye surgeons work through a microscope in an operation to remove a cataract.

kinds dramatically improved. Cyclosporin, which was isolated from a mold in 1976, was found to prevent the formation of the white blood cells that are responsible for rejection. Before the drug came into use, over half of all kidney transplants failed, and few heart-transplant patients lived more than two years after the operation. With cyclosporin, the success rate with kidneys rose to over 90 percent. Two-thirds of heart-transplant patients, who by the mid-1980s numbered several hundred each year in the United States alone, survived more than five years. Today, the figures are better still, and transplant surgery is a daily event.

Microscopes and keyholes

By its very nature, major surgery is a traumatic experience, particularly if it involves opening up the chest. In complete contrast, some forms of surgery developed during the 20th century depend on a delicate touch.

In 1921, C.O. Nylen, a Swedish ear specialist, became one of the first surgeons to carry out an operation with the aid of a microscope. He used a standard instrument with a single eyepiece, but in the years that followed, other eye and ear surgeons operated with the help of binocular microscopes, which give the user a sense of depth. Microscopes allowed an unprecedented degree of accuracy in the surgeon's work, enabling them to deal with structures too small or thin for surgery with the naked eye. These structures included the stapes, a minute stirrup-shaped bone in the inner ear; the cornea lying over the eye; and, in the 1960s, the nerves that operate fingers and toes.

After the Second World War, surgeons found that a delicate touch paid dividends in other areas, because the less a patient is injured by surgery, the faster he or she is likely to recover. At the beginning of the century, surgeons often made generous incisions to allow themselves room to maneuver. Now the trend moved in the other direction, with the emphasis on surgical techniques that involved minimal "invasion" of the body.

One way to achieve this is to avoid cutting into the body at all. An instrument called the

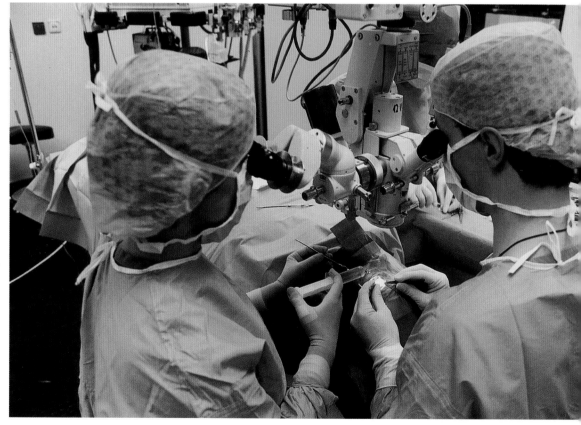

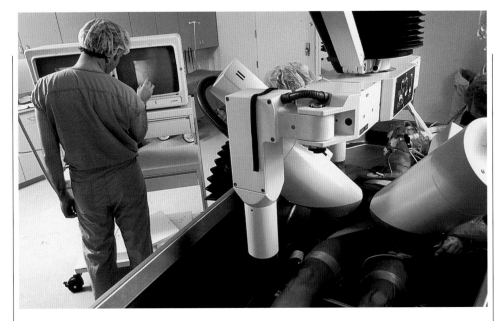

SHATTERED BY SOUND A patient floats in a water bath while being given ultrasound treatment to break up kidney stones.

lithotripter, first used in West Germany in 1980, has successfully replaced the scalpel in the treatment of calculi, or "stones." Calculi are crystalline mineral deposits that some-

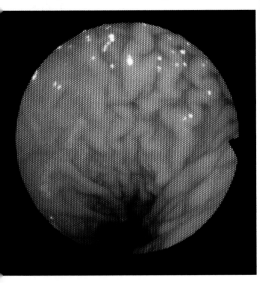

INSIDE STORY This internal view of the stomach was taken through a flexible gastroscope. This instrument can also be used to carry out minor surgery.

times form in the kidney, bladder and bile duct, which drains bile from the liver. Small calculi are harmless, but large ones can prevent organs working normally, and can be excruciatingly painful. Until the 1980s, the traditional method of treatment involved cutting open the organ concerned so that the stones could be extracted. The lithotripter removed the need for surgery: it works by focusing a beam of sound onto the stones, so that they shatter into pieces small enough to pass out of the body naturally.

The development of flexible endoscopes in the 1950s and early 1960s provided surgeons with new opportunities for minimizing the impact of operations. As well as being valuable in making a diagnosis, flexible endoscopes could be equipped with miniature surgical instruments and used to carry out operations, either through one of the body's natural openings or through a small incision or "keyhole."

Within a few years, "keyhole surgery" was being used to remove infected appendices, to repair hernias and to treat ulcers. Today they can even be used to remove gallbladders—an operation that previously required a major incision into the patient's abdomen. While transplant operations still attract much of the glamour attached to modern surgery, there is little doubt that these minimally invasive techniques have had an equally important impact.

REBUILDING THE BODY

In 1920, Harold Delf Gillies, a British surgeon born in New Zealand, published a landmark textbook called *Plastic Surgery of the Face*. It marked the emergence of a new specialization in surgery—the use of skin and bone as moldable or "plastic" substances, to re-create facial features damaged by severe accidents.

Gillies learned his skills during the First World War, when many soldiers were appallingly disfigured by combat during the long years of trench warfare, with its shell-borne shrapnel and subterranean mines. By taking small pieces of healthy bone and cutting it into small flakes, Gillies found he could rebuild the underlying framework of injured parts of the face. Using an instrument called a dermatome, he then shaved stamp-sized pieces of skin from the patient's arms or legs, and grafted these onto the rebuilt area. In the days before antibiotics, infection was a constant problem, but where it succeeded, plastic surgery helped severely injured soldiers to return to something approaching a normal life.

During the Second World War, Archibald MacIndoe, one of Gillies' colleagues and also a relative from New Zealand, helped to expand plastic surgery still further. One of his innovations was the pedicle graft—a skin graft that is initially nourished by a stem of tissue attached to its original site. In fact, pedicle grafts proved vulnerable to infection, but by 1945, MacIndoe's combination of artistry and technical skill had attracted international attention.

Gillies and MacIndoe both died in 1960, but by then their combined work had helped to make plastic surgery a part of mainstream medicine. Today, thanks to the discovery of a substance called epidermal growth factor, healthy skin can be shaved off a patient, grown in artificial conditions and then replaced after the underlying tissue has been reconstructed. As well as repairing damage and congenital abnormalities, it is also increasingly used in cosmetic surgery.

SPARE SKIN A piece of skin grown from a small number of donor cells is kept alive by a special nutrient jelly.

TECHNOLOGY AND THE BODY

MACHINES AND REPLACEMENT PARTS CAN NOW IMPROVE OR EVEN MAINTAIN LIFE IF THE BODY'S OWN ORGANS FAIL

In the early 1960s, President John F. Kennedy famously committed America to putting a man on the Moon by the end of the decade. Somewhat less famous, but almost as ambitious, was an engineering effort of a very different kind. Launched by the National Institutes of Health, it was intended to reach its climax in 1970 with the implantation of the world's first artificial human heart. In one version of the plan, the date earmarked for the operation could not have been more significant: February 14—St. Valentine's Day.

In the "can-do" atmosphere of the 1960s, with world economies booming, the artificial heart program formed part of a persistent theme in postwar medicine: the use of technology to support failing organs or body parts, or if necessary, to replace them altogether.

The kidney machine

More than 20 years before the artificial heart plan was launched, one of its most prominent figures, Willem Kolff, was honing his skills on a prior but equally significant tech-

nological challenge. Working in Holland in the 1940s while the country was occupied by the Germans, Kolff was developing a machine that would temporarily take over the function of a patient's kidneys in cases of

severe infection or poisoning. The kidneys play a central role in the chemical balancing act by which the body maintains its stable internal state. Their main contribution to this process lies in removing waste products as they build up in the blood, and adjusting the body's water balance. They do this by allowing waste products and water to diffuse out of the blood, before recapturing most of the water and returning it to the bloodstream. The concentrated waste forms urine, which the body can then dispose of.

This filtering process may sound simple, but the structures involved are extremely complex. A kidney contains about a million

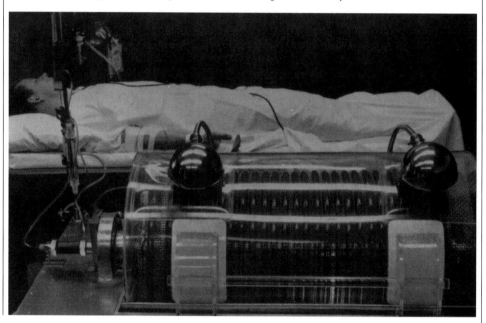

filtering units packed side by side, together with many miles of narrow blood vessels. Together they have a huge surface area through which selective filtration, or dialysis, can take place. To make an artificial kidney, this surface has to be replaced by a man-made

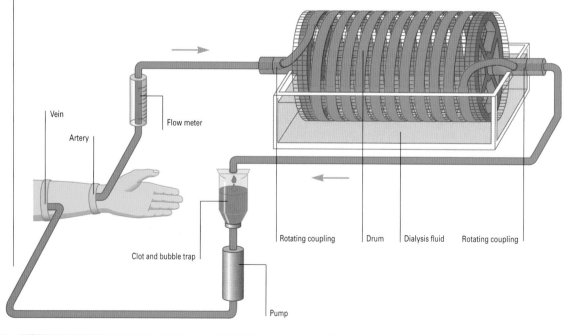

HOW DIALYSIS WORKS In this early kidney machine (above and left), blood flows through permeable tubing wrapped around a rotating drum. As the drum turns, waste flows from the blood through the tubing and into the dialysis fluid.

Vein

Artery

Flow meter

Clot and bubble trap

Rotating coupling

Drum

Dialysis fluid

Rotating coupling

Pump

1901 First hearing aid introduced

1930s First attempts to override faulty pacemaker in heart

1940s First attempts to renew hip joints

1945 First prototype kidney machine

membrane that can carry out the same work. The membrane must have pores to allow waste products to escape from the blood, but its pores must be small enough to block the path of the blood's essential proteins.

In 1913, experiments carried out at Johns Hopkins University had shown that a substance called collodion seemed to meet these requirements. The experiments also revealed that something was needed to prevent the blood clotting, or else the tubes carrying the blood would become hopelessly clogged up. The Johns Hopkins researchers solved the clotting problem in a bizarre way—by grinding up the heads of several thousand blood-sucking leeches. Leeches secrete an anticoagulant when they feed, and this substance, called hirudin, helped to prevent clots forming in the laboratory apparatus.

Kolff, fortunately, had access to a purified anticoagulant, which did away with the need to grind up leeches. However, in wartime Holland collodion was not available, so he was forced to look for something else. Instead, he used an alternative material with a similar chemical and physical structure—synthetic sausage skin. He realized from the outset that to achieve the same rate of dialysis as a living kidney, his artificial counterpart would have to be much larger than the real thing. As things turned out, when he completed the first prototype in 1945, his machine was almost as large as his patients.

Despite the difficult conditions surrounding its birth, Kolff's machine was an immediate success. He tried it out on people suffering from acute kidney failure, and discovered that it provided rapid relief from uremia, the condition caused by the build-up of toxic waste in the blood. After a few hours of dialysis, patients who were previously gravely ill could sit up and talk as if little were the matter with them.

Once it had proved its worth, Kolff's machine quickly spawned a wave of successors. In Kolff's original design, blood flowed through the sausage-skin tubing under the pull of gravity, but in many of the machines that followed in the late 1940s and 1950s—and indeed the ones in use today—the blood

was put under pressure. This helps to force out surplus water, and also speeds up the dialysis process. Kidney machines also shrank, although not by as much as their users might have wished.

When Kolff introduced his machine, it was intended to offer short-term relief to people who had suffered reversible kidney failure. But kidney machines proved so effective that they were soon being used to treat people who had no long-term prospect of a cure. In the 1950s and early 1960s, when kidney transplants were still in their

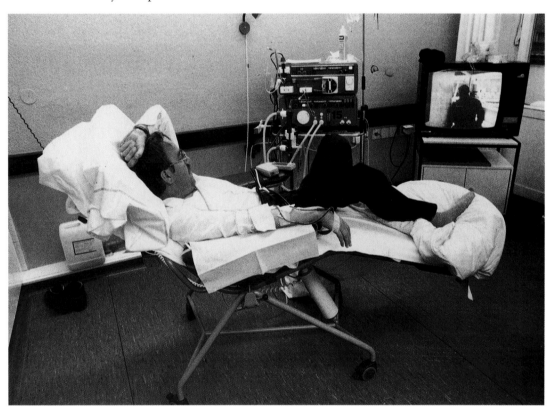

infancy, regular dialysis offered a new way to achieve something like a normal life.

This development raised some problems, one of which was the difficulty of connecting the patient to the machine. Originally, dialysis patients were linked to kidney machines by two surgical cannulas, or fine tubes, one inserted into an artery, and one into a vein. Repeated several times a week, this method of connection could soon run into trouble. The answer, an artificial shunt, was developed in the early 1960s. A shunt consists of a short semicircular tube, exposed outside the body, which artificially connects an artery and a vein in the wrist or lower leg. The shunt acts as a permanent "plumbing point" for linkage to the machine.

With the help of shunts and more compact machines, dialysis moved away from the hospital and into local clinics, and finally into patients' homes. Today, a small percentage of dialysis patients use kidney machines unaided, but many more use a much simpler form of treatment, called peritoneal dialysis, which was developed in the early 1970s. In peritoneal dialysis, sterile fluid is passed through a tube into the patient's abdomen, and a natural lining called the peritoneal membrane allows waste to filter into the fluid from the blood.

HOOKED UP Artificial dialysis is a long process. For this patient, television helps to pass the time.

The fluid is then allowed to drain out, taking the waste products with it. The entire process takes about an hour, and is usually repeated once a day.

Heart valves

In 1952, at a time when the kidney machine was undergoing a period of rapid development, another kind of technological aid had its first test on a human being. Charles Hufnagel, an American surgeon, inserted an artificial valve into a patient's aorta—the 1-inch-wide artery that carries blood from the

1952 First artificial valve inserted into a human aorta

1958 Fully implantable pacemaker produced in Sweden

1969 First artificial heart implanted into a human

1970s Peritoneal dialysis developed

1982 Barney Clark receives Jarvik artificial heart

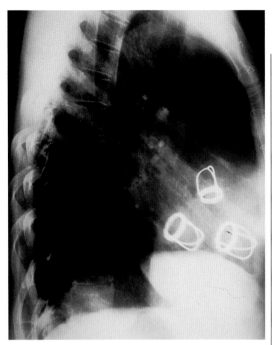

SEEING INSIDE THE HEART An X-ray shows a heart with artificial aortic, mitral and tricuspid valves.

heart to the trunk and legs. The valve was implanted to tackle a condition in which a faulty heart valve allowed blood to leak backward after each heartbeat.

Compared to Kolff's kidney machine, Hufnagel's device represented mechanics at its very simplest. It consisted of a plastic tube, topped by its one moving part—a ball trapped in an open wire cage. With each heartbeat, the ball moved away from the tube, allowing blood to flow past. Between heartbeats, when the blood tried to flow back, the ball was pushed against the tube, sealing it and stopping the flow. Although its application was quite new, its design was not: a remarkably similar object was granted an American patent in 1858. However, instead of being a heart valve, it was a humble bottle stopper.

Hufnagel's valve implant marked the start of a tremendous amount of research into the technology of heart-valve replacements. Structurally, the heart's four valves are extremely simple, which helps to explain how they can open and close 100,000 times a day without leaking and without getting jammed open or shut. But as heart surgeons soon discovered, man-made valve replacements

are not easy to produce. Natural valves have flexible flaps made of living tissue that can meet to form a blood-tight seal. Plastic flaps, which were tested in the early 1960s, initially seemed promising, but tended to lose their flexibility within a few months, and often had to be replaced. Artificial valves had some other worrying habits. One was that they produced blood clots, a problem that was tackled by giving recipients a steady supply of anticoagulant drugs. They also tended to destroy red blood cells, either by creating turbulence, or by trapping them when they shut. To make matters more difficult still, living tissue often grew into them, obstructing their openings or reducing the space inside them.

Some of these problems were overcome by using new materials, such as Teflon and Mylar, which let blood flow past with the minimum of friction. Others were tackled by making small but crucial adjustments to valve design. For example, the ball valve pioneered by Hufnagel was improved in the 1960s by a

HEART VALVES An artificial valve in the mitral position (right) prevents blood flowing from the left ventricle back into the left atrium. Tilting disc valves (below) snap shut if blood flows in the wrong direction.

heart surgeon, Albert Starr, and a retired aircraft engineer, Lowell Edwards, but it still suffered from the problem of narrowing through tissue overgrowth. Starr and Edwards overcame the problem by making the ball out of metal alloy rather than plastic, and by giving the valve's mouth a sharp edge. When the ball rolled back to shut the valve, it met the sharp edge and chopped away any tissue that had grown there.

Variations on the Starr-Edwards valve are still in use today, but they have been joined by several other designs that feature tilting discs instead of a moving ball. Together with

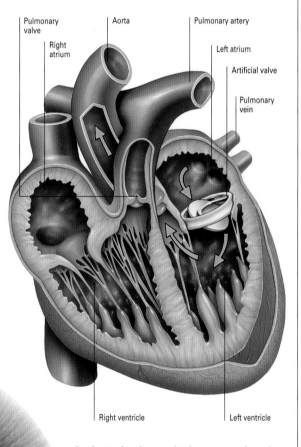

Pulmonary valve · Right atrium · Aorta · Pulmonary artery · Left atrium · Artificial valve · Pulmonary vein · Right ventricle · Left ventricle

biological valves, which come either from human donors or from animals, they have given many heart-disease sufferers a new lease on life.

Pacemakers

Unlike other kinds of muscle in the body, heart muscle triggers itself into action. A small area of the heart called the pacemaker initiates each heartbeat, and the chambers of the heart then contract in a rhythmic sequence. This, at least, is what happens in a healthy heart. If the pacemaker

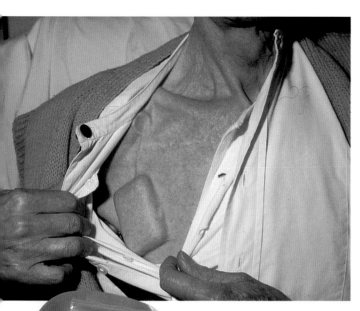

ON-BOARD TIMER Modern pacemakers are usually implanted beneath the collarbone, with wires leading through a vein and into the heart.

is defective, the rhythm can break down, leaving the body dangerously short of oxygen.

The first attempts to override a faulty pacemaker date back to the early 1930s, when Albert Hyman, a doctor in New York, built a hand-cranked machine that could deliver mild shocks at evenly spaced intervals. The machine's spring ran for about six minutes, often long enough to bring an irregularly beating heart back into line. Hyman's invention fitted into a doctor's bag, but as it needed winding up regularly it could only be used for short periods, and the patient had to keep still.

By the early 1940s, Hyman's original idea had already come a long way. The clockwork motor had been replaced with electrical components, which meant that pacemaking could continue for hours or days at a time. But despite this change, pacemakers from this period were still cumbersome. The power source had to be plugged in at a wall socket, and was often mounted on a trolley; and as in Hyman's machine, only the electrodes were implanted inside the body itself.

In 1952, surgeon Paul Zoll at Boston's Beth Israel Hospital used an external pacemaker to shock a man's heart back into action. For over two days the pacemaker alone kept the man alive, until his heart eventually recovered its own rhythm. Zoll's pacemaker was designed strictly for hospital use, but by the early 1950s a recent technological breakthrough—the invention of the transistor—was beginning to revolutionize pacemaker design. Instead of being the size of a car battery, pacemakers were rapidly shrinking. At the end of the decade decoupled pacemakers appeared, with a small external powerpack that operated internal electrodes by radio waves, removing the need for wires that stuck out through the skin. From here the ultimate goal was in sight: a pacemaker complete with its own power supply that could be permanently implanted in the body.

On both sides of the Atlantic, research teams closed in on this objective. In 1958, the Swedish engineer Rune Elmqvist produced a fully implantable pacemaker, and in 1960 Åke Senning inserted it into a man's abdomen, with wires leading up to his heart. Unfortunately, the implant failed after just three hours, and its replacement lasted just eight days. However, the patient remained undaunted. He went on to become an enthusiastic member of the Swedish research team, "test driving" more than 20 different implants over the following two decades.

Since the 1960s, new power sources—including pellets of radioactive "fuel"—have meant that pacemakers can be left in place for years without needing any maintenance. With the arrival of microchips, pacemakers have also become much more versatile. Instead of sending out shocks at the same fixed interval, regardless of how much work the body is doing, they can now adjust to the body's changing needs, when resting or exercising. Some pacemakers also monitor the heart's performance, and step in immediately if the heart shows signs of fibrillation—a form of fast, fluttering heartbeat that can be fatal. Within a few seconds, the heart is gently nudged back into step.

Artificial hearts

Heart valves and pacemakers are now an everyday part of medical technology and practice, and are fitted by the thousand
continued on page 82

ARTIFICIAL HEART The Lepare artificial heart is used as a temporary measure until a heart transplant can be performed.

A TRIP TO THE DENTIST

THANKS TO IMPROVED ANESTHETICS AND HIGH-SPEED DRILLS, DENTAL TREATMENT HAS SHED THE MIXED REPUTATION IT HAD AT THE START OF THE CENTURY.

In the early years of the 20th century, a trip to the dentist was a distinctly uncomfortable experience. Dentists had local anesthetics—the most useful, novocaine, was discovered by a German chemist in 1904—but by modern standards, dental technology was crude. The era of pedal-powered machinery had come to an end in the late 1800s, after being ushered out by electricity. But the primitive drills of the time still made the salvaging of diseased teeth difficult: in many cases, the easiest solution was to remove them altogether.

Because many people lost their teeth at a relatively young age, the manufacture of dentures and crowns was a flourishing business. At the beginning of the century, denture bases were made of vulcanite, a substance produced from natural rubber. Vulcanite had been used in dentistry since the 1880s, but in 1919, largely on cosmetic grounds, it was replaced by pink denture rubber. In the 1930s this in turn was superseded by vinylite, and then by resin-based plastics.

The typical dentist's office of 1900 usually served as an administrative space as well. By modern standards, it contained a strange jumble of domestic furniture and medical equipment, with few concessions to the need for easy cleaning. The chair was often positioned close to a large window—a legacy of the days before electric lighting—and was designed to be used with the patient in an upright position. At this time dentists, who were still almost all men, did their work standing up and leaning over their patients. The damaging effects of this unhealthy posture were revealed many decades later, when the design of dentists' chairs changed sufficiently to allow dentists to work sitting down. Remarkably, this extended the average dentist's life expectancy by about 15 percent.

Gas containers were a standard feature in dentists' offices in 1900, but X-ray equipment did not become

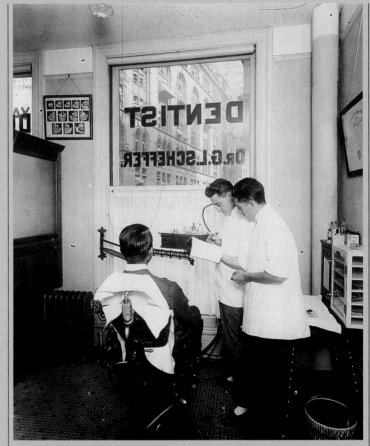

QUICK CHECK An American dentist consults a patient's records in the early 1900s. In this office, work was carried out entirely by daylight.

widespread until after the First World War. Typical dental X-rays were about the size of a postage stamp, and showed just a small part of the jaw. Panoramic X-rays, which show all the teeth in both jaws on one piece of film, did not appear until much later.

While local anesthetics or gas dealt with the pain of treatment itself, early 20th-century dentists did not have any drugs that could help to sedate severely anxious patients. Real progress in this field did not come until the early 1960s, with the development of benzodiazepines. These drugs, which include Valium, found their way into dental practice from about 1966, and are still used today. Until the 1940s, dentists also lacked antibiotics—something that made severe infections much more difficult to treat. Without antibiotics,

LASTING IMPRESSION Dental casts, made from quick-setting plaster of Paris, were originally produced using latex molds. Today synthetic molds have taken their place.

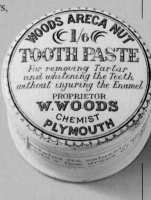

POWDERS AND PASTES Before the invention of the squeezable tube, toothpaste was sold in earthenware pots.

abscessed teeth could be highly dangerous, because there was always a risk that they could lead to blood poisoning or septicemia.

Perhaps the greatest improvement in dental treatment came about in the 1940s, with a complete redesign of the dental drill. At this time, drills were driven by a system of belts and gears, and were clumsy and difficult to control. John Walsh, a dentist working in Australia, realized that because they turned relatively slowly, the dentist had to press hard to make them work. This pressure sent disagreeable vibrations through the patient's teeth and skull.

After carrying out some experiments with tuning forks, Walsh tried out faster drills that produced higher-frequency vibrations. He managed to raise their speed from about 1,000 rpm to 6,000 rpm, and found that they were better for both dentist and patient. By the early 1950s, Walsh's ideas had been taken up in America and Europe, and new fast drills started to appear in many dentists' offices. However, there was a snag: at very high speeds, drilled

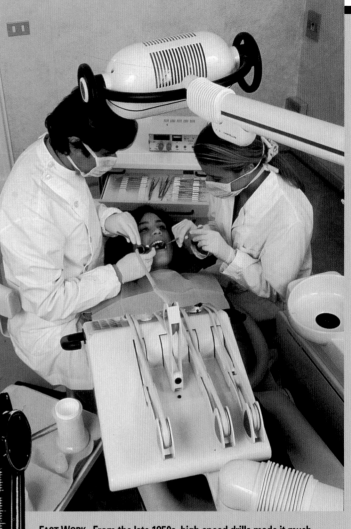

FAST WORK From the late 1950s, high-speed drills made it much easier for dentists to carry out precision work.

PANORAMIC VIEW A dental nurse lines up a panoramic X-ray gun in the 1930s. For the first time, this new piece of equipment revealed all a patient's teeth in a single picture.

teeth became so hot that they started to burn. The answer was to spray water on the tooth while it was being drilled, adding suction units to the growing list of dental equipment.

The first high-speed drills were still driven by belts and gears, but in the late 1950s, gearless turbine drills came into use. Driven by water pressure and later by air, they achieved unprecedented speeds and halved the time needed for most procedures. The Borden Airotor, for example, which was one of the first air-driven models, could reach 300,000 rpm. It turned faster than the turbine in a jet engine, and gave off a high-pitched whine. This sound, which was quite unfamiliar to patients in the 1950s, has since become the dentist's trademark.

The postwar era also saw improvements in preventative action. In 1945, after lengthy clinical trials, fluoridation of the water supply began in the United States, and was soon followed in other countries. Fluoridation has generated controversy because it is a form of imposed medication, but it has produced valuable results. A 20-year study, carried out in Illinois and published in 1967, showed that the addition of fluoride to the public water supply had reduced tooth decay by over 50 percent.

HELP WITH HEARING

The world's first electrically amplified hearing aid was introduced in 1901. Early versions were large and cumbersome, but with the invention of the transistor in the 1940s they shrank rapidly. Today's in-the-ear models use alkaline batteries and microelectronics to pack the same amplification power into a much smaller volume.

HEARING AIDS In the 1930s, hearing aids were large and obtrusive (below). Today's models (right) are small enough to wear tucked discreetly behind the ear.

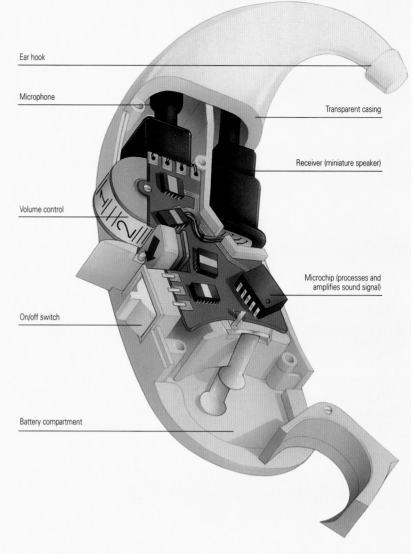

Ear hook

Microphone

Transparent casing

Receiver (miniature speaker)

Volume control

Microchip (processes and amplifies sound signal)

On/off switch

Battery compartment

every year. However, the same cannot yet be said of the most complex implantable organ so far designed: the artificial human heart.

In the mid-1960s, when the American artificial heart program began, the arguments for it looked persuasive. According to a report released in 1966, once artificial hearts went into mass production, surgeons in the United States alone could be expected to use more than 100,000 a year. With sick people returned to active life, the economic benefits could well outweigh the cost of research. The program went ahead, and on April 4, 1969—three months before humans first set foot on the Moon—the surgeon Denton Cooley implanted an artificial heart driven by

PARTIAL PUMP A left ventricular assist device (LVAD) is sometimes fitted to patients with failing hearts who are waiting for transplants.

compressed air into Haskell Karp, a man suffering from severe heart failure. The artificial heart was a temporary measure, and three days later Karp was given a heart from a donor. However, within a day of the second operation, the patient died.

Cooley's bold move generated a huge amount of publicity, but did not do much to inspire confidence in the artificial heart that he used. Instead, attention moved to the University of Utah, where Willem Kolff—inventor of the kidney machine—was carrying out his own research into artificial organs. In 1971, Kolff engaged an assistant named Robert Jarvik, with the express aim of perfecting an artificial heart.

During the 1970s, Jarvik tested a series of prototypes on animals. Jarvik hearts, as they became known, were made from polyurethane and aluminum, and were powered either by compressed air or electricity. Initially, Jarvik's experimental animals lived for just a few days, but by 1979, the Jarvik-7 model achieved a record of seven months. Willem Kolff decided that the time had now come to use a Jarvik heart on a human being. On December 2, 1982, at the University of Utah, Barney Clark, a 61-year-old dentist from Seattle, was given a Jarvik-7 heart amid a blaze of publicity. Clark survived the operation, but after several weeks developed kidney disease, intestinal ulcers and a severe lung infection. He died 112 days after the operation took place, and the bill for his care alone was estimated at more than $250,000. Some surgeons considered it a worthwhile experiment, but to Jarvik's critics Clark's experience was not so much a prolonging of life as a prolonging of death.

Since 1982, fewer than 100 artificial hearts have been implanted into human patients, with the longest survivor living just 20 months. Poor performance and even poorer quality of life have led to artificial

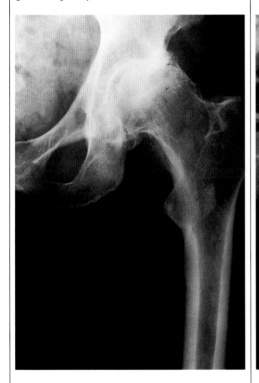

heart implantation being banned in many countries, including the United States. However, the idea of mechanically aiding the heart has not faded away. Instead of replacing the heart, research is now focused on mechanisms that work alongside it.

One of these heart-helpers is the left ventricular assist device (LVAD). This externally powered pump helps the left ventricle, which is the largest chamber of the heart, with the most work to do. The LVAD takes blood from the left ventricle and pumps it into the aorta, lightening the load on the heart itself. Another device is the hemopump, a bullet-shaped turbine introduced in 1989 that can be worked up the body from an artery in the groin, until it lodges next to the heart. Here the blades of the turbine suck blood out of the heart, helping to pump it around the body. Neither of these devices cures the problem of an ailing heart, but for a patient waiting for a heart transplant, the extra time they buy can mean the difference between life and death.

Joints and lenses

Medical technology is not only used to save life, but also to improve it. As many millions of people have discovered, two of the greatest 20th-century successes have been the replacement of limbs and failing joints and the development of artificial lenses.

Replacement joints are not a recent invention, but for a long time they were limited by a lack of materials. Nineteenth-century doctors had some success with external replacement elbows, but despite some experimental surgery, replacement of the hip joint proved impossible. Unlike the elbow, the hip has to bear extremely strong forces acting on it while being able to move in almost any direction. In the 1940s, the American surgeon Marcus Smith-Petersen made the first modern attempt to renew hip joints affected by arthritis. He lined the hip bone's socket with a cup made of polished metal, and reduced the ball-shaped head of the femur, or thighbone, so that it would fit in the new socket. The new joint soon ran into trouble, however, because the femur's head often dislocated, or jumped out of position. Even with the femur in place, the joint was much stiffer than the real thing.

In the 1950s, two French surgeons, J. and R. Judet, took a more radical approach. They removed the corroded head of the femur entirely and replaced it with a plastic hemisphere attached to a spike. The spike was driven into the soft core of the femur, where—in theory at least—a strong cement would lock it in place. This joint also encountered difficulties. The plastic head was not slippery enough, which put strain on the spike. Eventually the entire plastic replacement worked loose, rendering it useless. A combination of plastic and metal did eventually provide the answer, however.

In 1960, John Charnley, an orthopedic surgeon at the Manchester Royal Infirmary in England, reversed the Judets' approach. He tried a metal ball

REPLACEMENT HIPS X-rays show how an artificial hip joint mimics the bones it replaces. The ball of this artificial joint (left) has a metal-ceramic coating to reduce wear and give the joint extra strength.

LIFE ON HOLD

MODERN OPEN-HEART SURGERY IS ONLY POSSIBLE THANKS TO A DEVICE THAT KEEPS THE BODY ALIVE WHILE THE HEART IS BROUGHT TO A HALT

One of the most important developments in medical technology—the heart-lung machine —had its practical origins in the mid-1930s, with the work of an unlikely sounding team. One partner was Alexis Carrel, the French pioneer of blood-vessel surgery, who had left his native France for the United States. The other was Charles Lindbergh, the American aviator and national hero who made the first nonstop crossing of the Atlantic. Carrel's interest in human circulation was strictly professional, while Lindbergh's was more personal: his sister-in-law had a serious heart condition, which he thought technology might be able to improve.

In 1935, Carrel and Lindbergh demonstrated their brainchild—an external heart, or

SUPPORTING ROLE   A heart-lung machine is used during a coronary artery bypass operation (left). The heart-lung machine (HLM) maintains blood flow, oxygenates the blood and removes carbon dioxide, and controls body temperature (below).

perfusion pump, which could circulate the blood. It was never tested on humans, but it did show the feasibility of maintaining circulation by artificial means. Other researchers had also discovered that blood could be artificially oxygenated, but at the time no one managed to create a machine that would carry out both tasks at once.

Four years before Carrel and Lindbergh perfected their pump, John Gibbon, an American surgeon, had watched a patient struggle for life after an emergency operation. "During that long night's vigil . . . the thought naturally occurred to me that the patient's life might be saved if some of the blue blood in her veins could be continuously withdrawn into an extracorporeal blood circuit, exposed to an atmosphere of oxygen, and then returned to the patient by way of a systemic artery. . . ." In his imagination, Gibbon had created a heart-lung machine. All that remained was to build it.

Gibbon's work took over two decades, and was dogged by the problem of oxygenating enough blood to keep an adult alive. During the Christmas of 1946 he took his existing designs to the head of research at IBM, who agreed to help. By 1952, the problems of oxygenation had been overcome, and the heart-lung machine was ready for use. The first patient, a year-old child, died soon after the operation from an unforeseen complication. The second, an 18-year-old girl, was kept alive by the machine for 26 minutes while a hole between two of her heart chambers was sealed. She later made a complete recovery.

The heart-lung machine is often used in conjunction with hypothermia, a medical technique pioneered in the 1940s. During hypothermia, the body's temperature is lowered from its normal level of 98.6° Fahrenheit to about 86° Fahrenheit, which reduces the patient's need for oxygen and the work that the machine has to do.

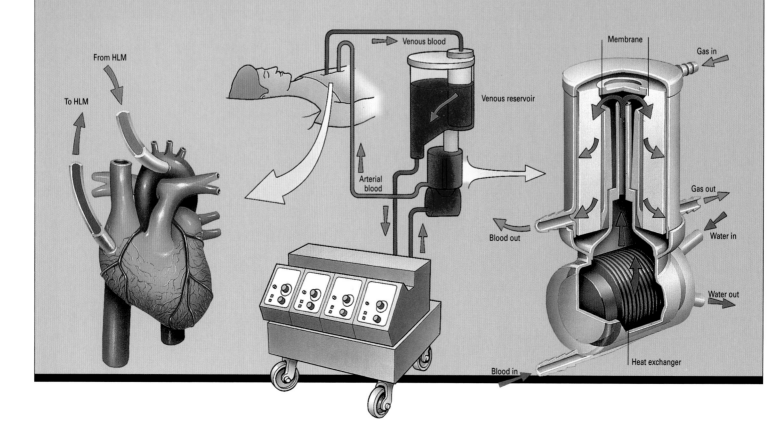

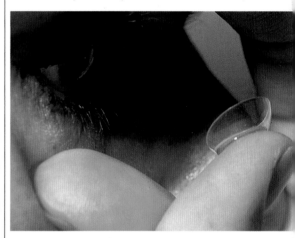

HIGH-TECH HAND Robotics technology gives a prosthetic hand added flexibility and articulation.

and a plastic socket. His original choice of plastic was Teflon, but this quickly wore out. He replaced it with a form of high-density polythene and achieved his goal—an artificial joint that produced almost as little friction as the real thing. A new cement called methyl methacrylate glued ball and socket in place. Since 1961, when Charnley reported his new technique, hip replacement has become so common that it excites little attention, and many other major joints—including the knees, shoulders and elbows—can also be replaced.

Plastic has also played a key role in devices to improve vision. One of these, the implantable plastic lens, has been used since the 1950s to restore sight to people whose own lenses have become clouded by cataracts. Cataracts develop when the proteins in the eye's lens deteriorate with age, turning the lens milky.

Plastic has revolutionized contact lenses, which until the late 1930s were made of glass. Glass contact lenses were originally designed to cover most of the eye, and they were so uncomfortable that few people chose to wear them. At first, plastic lenses were not much better, although things improved with the development of corneal lenses in 1948, which only cover the domed part of the eye in front of the pupil. The problem with both glass and plastic lenses was that they were made of hard, impermeable materials. They prevented oxygen from reaching the cornea, and stopped tears washing the eye in the normal way.

In 1960, a Czech chemist, Otto Wichterle, devised a new kind of lens based on a soft, porous plastic that could absorb water, and by the early 1970s, contact lenses

SOFT FOCUS A contact-lens wearer prepares to put one in place. Today's soft lenses are comfortable enough to wear all day long.

made of this plastic started to tempt many people away from wearing glasses. Although they are extremely comfortable, these lenses do have one major drawback— they absorb proteins and bacteria from the surface of the eye, which means that they have to be kept scrupulously clean. In the 1970s, lens hygiene was a laborious process, involving protein-removers and disinfectants, but with the increasing popularity of soft contact lenses, things began to change. Mass production meant that the price of soft lenses plunged, turning them into disposable items. Today, soft lenses are worn by millions of people, and are thrown away without a second thought.

COMPUTERS AND COMMUNICATION

The world's first electronic computer was built in 1946: it contained more than 19,000 valves and weighed 30 tons. With the invention of the transistor in 1948, however, and the microchip in the 1970s, computers shrank, and began to be used in medicine. In addition to providing imaging techniques and radiotherapy treatment, computers have found their way into a new generation of aids for the disabled, helping people lead more independent lives.

Unlike mechanical machines, electronic circuits can be activated in several ways. Since the early 1970s, medical engineers have created a range of devices that can be operated by the slightest movements. For example, voice synthesizers can be activated by the touch of a finger or even by the movement of an eye, while electronic hands can be operated by changes in pressure created by flexing muscles in the arm. In the future it seems likely that these artificial aids will be even more precisely controlled, by connecting them directly to nerves.

Computer technology has given many severely disabled people a chance to express themselves in a way that previously would have been impossible. Most famously, perhaps, a voice synthesizer has enabled the British physicist Stephen Hawking, author of *A Brief History of Time*, to continue his writing and research, despite being immobilized by motor neuron disease.

LET'S TALK A disabled boy wears a voice-synthesizer that he controls with a laser fitted in his headband.

ALTERNATIVE MEDICINE

DISSATISFACTION WITH MODERN MEDICINE HAS FUELED A REVIVAL IN ALTERNATIVE TECHNIQUES—SOME NEW, OTHERS CENTURIES OLD

During the 20th century, science was used to explain everything from the origin of the universe to the common cold. Using scientific logic, doctors grew accustomed to reducing medical problems to bare and often microscopic essentials—such as attacks by bacteria or viruses, or faults in the way the body's cells work. Once the cause of the problem was pinned down, science could offer treatment in the form of drugs, surgery, or some other type of intervention.

In several areas of human health, this science-based approach to medicine has proved outstandingly successful. In 1900, the average life expectancy in the developed world was only about 46 years, mainly because so many children died in their early years. But with a scientific approach to public hygiene, and new ways of combating disease, childhood death rates plunged. Today, average life expectancy has risen to more than 70 years in developed countries, and in Japan—the country with the highest life expectancy of all—it currently stands at about 82 years for women and 76 years for men. The drop in childhood disease is only part of the story. Death rates from cancer have remained much

TRADITIONAL TREATMENT In Japan sand-bathing (right) uses heat and in China cupping (below) uses suction to improve health.

the same, but those for many other diseases of adulthood—such as strokes and diabetes—have also fallen.

Despite these immense achievements, the 20th century—and its latter years in particular—saw a growing discontent with the way modern medicine works. With health care growing ever more complex and impersonal, many people turned away from it in their search for well-being. Instead, they turned toward approaches that address the whole person, rather than just the symptoms of a specific disease as orthodox Western medicine does. The result has been a huge surge of interest in alternative therapies—forms of treatment that scientific medicine once sought to stamp out altogether.

Culture clash

Until the 1960s, orthodox medicine held a virtual monopoly in dealing with illness and disease throughout the Western world. With several decades of rapid development behind it, particularly in the fields of surgery and pharmacology, it had an overwhelming authority and a seemingly unstoppable momentum. The white-coated doctor was a symbol of this technical progress—a person who had scientific knowledge at his or her fingertips, and whose time was so precious that the average patient could be spared only a few minutes.

This medical monopoly was reinforced by law and by professional regulations. In many western European countries, most forms of alternative medicine were technically illegal, although the law was not always rigorously enforced. In Britain, doctors could be struck off the medical register simply for referring patients to anyone who was not medically qualified, and in the United States, some advocates of alternative approaches to medicine even ended up in prison. To emphasize its view that they were outsiders, the medical establishment dismissed such people as being

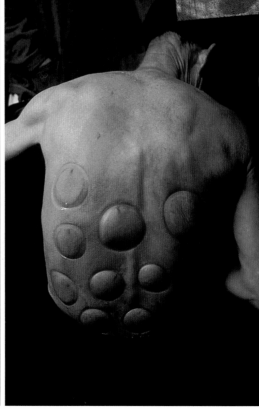

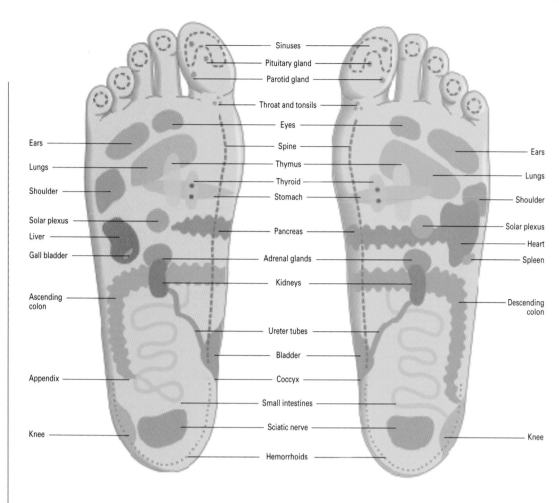

- Sinuses
- Pituitary gland
- Parotid gland
- Throat and tonsils
- Eyes
- Spine
- Thymus
- Thyroid
- Stomach
- Pancreas
- Adrenal glands
- Kidneys
- Ureter tubes
- Bladder
- Coccyx
- Small intestines
- Sciatic nerve
- Hemorrhoids

Ears
Lungs
Shoulder
Solar plexus
Liver
Gall bladder
Ascending colon
Appendix
Knee

Ears
Lungs
Shoulder
Solar plexus
Heart
Spleen
Descending colon
Knee

FEET FIRST In reflexology, discomfort in the foot is related to disorders in specific parts of the body (left). A reflexologist treats a patient by giving compression-massage in the relevant area of the foot.

"unorthodox," "unconventional," or belonging to the "medical fringe"—words that were carefully chosen to indicate disapproval. Science held the key to medical progress, and all else was bound to fail.

As the 1960s began, two things happened to upset this state of affairs. The first was a change in patients themselves. Unlike preceding generations, the generation that grew up in the 1960s was rebellious and individualistic, and much less inclined to accept the dictates of those in positions of authority, including members of the medical profession. At the same time, some of the weaknesses of scientific medicine were beginning to make themselves apparent. Medicine had conquered many of the epidemics that ran riot in earlier decades, but people still fell ill. Instead of preventing disease, medicine seemed merely to have shifted ill health from one group of diseases to another.

The statistics make surprising reading. In the United States, for example, the average patient questioned in the 1920s reported

SPIRITUAL HEALING Healing through the use of spiritual energy—seen here at a commune in Poona, India—is one of the oldest forms of alternative medicine.

about one bout of acute disabling illness a year, lasting on average 16 days. By the early 1980s, the reported illness rate had doubled, and the duration had increased to 19 days. The number of people who described themselves as satisfied with their health and physical condition also fell—a finding reflected in other countries. On paper, people should have been feeling healthier but, subjectively, they were feeling the reverse.

Arthur Barsky, an American psychiatrist who correlated this information in the 1980s, described it as the "paradox of health." He recognized that there were—and are—several explanations for this phenomenon. One is that during the 20th century, people became much more interested and informed on the subject of health. A greater knowl-

1956 German alternative therapists resume work, having been banned by the Nazis

1970s American doctors invited to China to see traditional medical practices

1988 Controversial paper on homeopathy published by *Nature*

edge of medical matters, and the problems that can afflict us, seems to have eroded people's sense of well-being. Another explanation lies in the nature of medical care itself. In an era in which doctors and surgeons have pursued ever more dramatic forms of medical intervention, expectations have risen. But with this has come an important side effect: when those high expectations are not fulfilled, the result is frustration,

WILHELM REICH AND THE ORGONE BOX

During the 1940s, the Austrian-born philosopher Wilhelm Reich developed a novel form of therapy called bioenergetics, which was based on "orgone," a vital sexually-based force that Reich believed permeated the universe. Working in the United States, Reich devised a box-like instrument called an "orgone accumulator" to gather this energy for therapeutic use. Orgone therapy attracted many adherents, but the press was scathing, and he eventually ran afoul of the law. The Food and Drug Administration declared that orgone did not exist, and impounded his stock of orgone accumulators, together with his research records. They were later burned in an incinerator on Long Island. Reich himself ended up with a two-year prison sentence, and he died in 1957 while still behind bars.

COURTING CONTROVERSY Wilhelm Reich in 1955, before his prison sentence.

disappointment, and a willingness to try something else.

Ancient and modern alternatives

In most Western countries alternative medicine was, until the 1960s, largely dismissed as quackery, reminiscent of times when unqualified physicians made a living by duping gullible clients. In Germany, a country with a strong tradition of "natural"

medicine, alternative therapists or *Heilpraktiker* were given freedom to resume their work in 1956 after being banned by the Nazis, but elsewhere alternative medicine attracted relatively little interest. However, one persistent exception to this was a form of medicine that has baffled and infuriated orthodox doctors throughout the century—homeopathy.

Homeopathy was devised by Samuel Hahnemann, a German doctor who died in 1843. Initially at least, Hahnemann's medical career followed orthodox lines, and he showed himself to be both competent and perceptive. During his studies he developed a particular interest in chemistry and pharmacology, and devised and sold his own medicines.

In the course of his work, Hahnemann investigated a long-standing medical principle that "like cures like"—in other words, that a drug that produces mild symptoms of a disease will also help to cure it. Arsenic, for example, produces stomach pain, so in accordance with the principle of like cures like, it was used as a medicine for stomach disorders, sometimes with catastrophic results. Hahnemann's innovation was to dilute the active ingredient so much that it could not possibly do any harm. But in diluting an ingredient up to 40 or 50 times in succession, he effectively reduced its concentration to zero.

According to science-based medicine, there is no possible physical mechanism by which such a dilute drug could work. However, since Hahnemann's time millions of people have taken homeopathic remedies,

LIKE CURES LIKE Homeopathic medicines sell in ever increasing quantities—despite the fact that many doctors consider them to be ineffective.

HEALING HERBS An herbalist sorts different types of herbs. Despite synthetic drugs, traditional herbal remedies are widely used.

and a large proportion have found that they have a beneficial effect. This challenges fundamental scientific laws, which is why homeopathy came under such vigorous attack for most of the 20th century.

One explanation repeatedly advanced by orthodox doctors is that homeopathy works through the placebo effect, meaning that its results are due to purely psychological changes in the patient. As doctors have discovered, this effect—which means "I shall please" in Latin—is remarkably powerful, and can easily be tested. For

HOMEOPATHY ON TRIAL

ACCLAIMED BY ITS ADHERENTS AND RIDICULED BY ITS CRITICS, HOMEOPATHY HAS GENERATED FIERCE CLASHES IN THE HEART OF THE SCIENTIFIC ESTABLISHMENT

On June 30, 1988, the science journal *Nature* published a research paper by Professor Jacques Benveniste, a leading French biochemist. The paper was preceded by the editorial equivalent of a health warning. Under the heading *When to believe the unbelievable*, the editor advised readers to treat Benveniste's paper with extreme caution, because it seemed to threaten the basis of science for more than two centuries.

Benveniste's research had a direct bearing on techniques used in homeopathy. He had prepared solutions of immunoglobulin E, which produces observable changes in white blood cells. The strongest solution contained 1 percent of the active ingredient, but the weakest had been diluted tenfold 120 successive times, until it was so weak that it was effectively pure water.

FANNING THE FLAMES   Professor Benveniste (left) suggested that a chemical mechanism might lie behind homeopathy.

he commented: "Readers of this article may share the incredulity of the many referees who have commented on several versions of it during the past several months...With the kind collaboration of Professor Benveniste, *Nature* has therefore arranged for independent investigators to observe repetitions of the experiments. A report of this investigation will appear shortly."

This ominous footnote heralded the arrival of a team of "ghostbusters" in Paris, led by Maddox himself. The experiments were repeated under video surveillance, with security codes being used to prevent different solutions being identified by anyone while the experiments were carried out. Benveniste described the experience as being like a Salem witch hunt, with the team going through his laboratory notebooks, and interrogating members of his staff.

On October 27, 1988, the team from *Nature* published their report. They identified flaws in Benveniste's laboratory procedures, and described his statistics as being too clear-cut to be true. Crucially, they failed to reproduce his results. However, Benveniste pointed out that laboratories in Canada, Italy and Israel had copied his experiments, and had come up with the same remarkable findings. Accusations and counter-accusations were exchanged, until finally the editor of *Nature* declared the matter closed.

The "Benveniste affair" still poses many puzzles, one of which is a strange wave-like pattern in Benveniste's results. But whether his work is a major breakthrough, or simply the result of some oversight in laboratory technique, no one yet knows.

SCIENTIFIC PAPER   NATURE VOL. 333 30 JUNE 1988

Human basophil degranulation triggered by very dilute antiserum against IgE

E. Davenas, F. Beauvais, J. Amara*, M. Oberbaum*, B. Robinzon†, A. Miadonna‡, A. Tedeschi‡, B. Pomeranz§, P. Fortner§, P. Belon, J. Sainte-Laudy, B. Poitevin & J. Benveniste‖

When Benveniste tested these solutions on white blood cells, he reported extraordinary findings. Instead of having a weaker effect, the highly dilute solutions often seemed just as active as the stronger ones. Even the weakest solution of all, which theoretically contained no immunoglobulin E, seemed to trigger changes that could be observed with a microscope. These changes occurred only if the solutions had been vigorously shaken—exactly the procedure that homeopaths use when preparing medicines. Benveniste advanced some tentative explanations for this effect, one of which was that water might act as a form of "template," preserving the form of the immunoglobulin molecules. However, he freely admitted that he was baffled by his findings.

Homeopaths immediately cited the research as evidence for the power of extremely dilute drugs. John Maddox, the editor of *Nature*, was not so impressed. In a footnote to Benveniste's paper,

HIT SQUAD   Members of *Nature* magazine's investigation team examine data gathered at Professor Benveniste's laboratories.

example, in 1929, W.E. Dixon, a British professor of physiology, carried out an experiment on the placebo effect by giving a group of men a dose of caffeine each day before they started work. Caffeine is a stimulant, and as expected, it made the men work harder. But when the morning's dose of caffeine was secretly replaced by another substance, which did not have a stimulant effect, there was no change in the amount of work the men did. They expected the drug to have a particular effect, and it did.

In later decades, research into the placebo effect revealed that it is extraordinarily far-reaching. Doctors found that the power of suggestion, conveyed in the ritual of a daily dose of placebo medicine, could sometimes achieve genuine cures. The concept of the body as a machine, entirely divorced from the mind, began to look increasingly shaky, and made a psychological explanation for homeopathy more and more plausible. But research into homeopathy produced a startling and inconvenient fact that could not be explained in this way: it seemed to work on animals as well.

During the 1980s, Professor Jacques Benveniste, a French biochemist, carried out some research that seemed to show that extremely dilute substances can produce physical effects. But Benveniste's work attracted an unprecedented storm of criticism and he was openly accused of fabricating results. Today, homeopathy is more popular than ever, but the world of orthodox medicine is still undecided how, why, or even if it works.

Looking East

During the late 1960s and early 1970s, millions of young people looked to the East for spiritual inspiration. The Beatles flew to India and took up transcendental meditation under the guidance of the Maharishi Mahesh Yogi, and eastern philosophies and religions—such as yoga and Buddhism—found new converts in the West. At the same

HEALING NEEDLES Acupuncture needles like these are inserted at specific points in the body.

time, Eastern medicine also traveled westward, bringing ancient forms of treatment to an increasingly interested public. One of these treatments was acupuncture.

Acupuncture involves sticking fine needles into precisely defined points on the body. The points are arranged in pathways called meridians, and each one relates to a particular organ or bodily function. In traditional acupuncture the needles are simply inserted and left in place, but in more modern forms they may be rapidly twirled, or moved up and down, or connected to a mild electric current. The effects claimed for acupuncture range from improvements in digestion and hearing to complete anesthesia in particular parts of the body.

Acupuncture has always been practiced in

movements of t'ai chi, carried out in the quiet period at the beginning of each day, encourage the free flow of *Qi*, the life-force that energizes all living things. If *Qi* is able

CHINESE HERBAL REMEDIES Many of the traditional remedies used in China are based on plants that are not found in other parts of the world.

to flow unimpeded, good health results.

At the outset of the long Chinese civil war of 1927-49, the ultimately victorious communists initially committed themselves to eliminating traditional medicine in favor of medicine based on science. However, during the course of the war there was a change of heart. In 1944, Mao Tse-tung announced that "to surrender to the old style is wrong; to abolish or discard is wrong." As a result of this directive, Chinese medicine survived the arrival of its Western equivalent, and one form began to complement the other.

In the 1970s, international politics helped propel acupuncture toward much greater worldwide prominence. When President Nixon embarked on a policy of opening diplomatic channels to China, American doctors were invited to China to see for themselves how the Chinese medical system worked under communist rule. During their travels they encountered "barefoot doctors"— members of an army of paramedics who took both traditional and Western medicine throughout the countryside, on foot or by bicycle. They also had a chance to witness acupuncture firsthand in clinics and operating rooms.

This resulted in a flurry of reports in med-

ical journals and an extraordinary wave of interest among the public at large. The reports themselves were often enthusiastic, although couched in scientific and cautious terms, with notes about the traditional Chinese stoicism in the face of pain, the implication being that American patients might not find acupuncture as effective as they imagined. But public interest did not wane, and acupuncture spread

IATROGENIC ILLNESS

Iatrogenic illness is a seeming contradiction in terms—an illness created by medical treatment. However, iatrogenic illnesses are as old as medicine itself, and with the advent of modern drugs and surgery, their potential for harm has increased. In 1967, a report into the hazards of hospitalization in America stated that one in five patients admitted to research hospitals acquired a treatment-related disease. In one in 30 cases, the disease proved fatal.

Chinese communities scattered across the world, without making any headway in Western culture. On an official level, it was introduced into Europe by French doctors in the 19th century, although at the time it failed to provoke much interest.

Unlike homeopathy, which was largely the creation of one man, acupuncture is part of a complex system of traditional medicine that has existed for several thousand years. Over the centuries, Chinese medicine has developed concepts of anatomy and physiology quite different from those that emerged later in Western medicine. It also evolved a wide range of herbal remedies, along with techniques such as t'ai chi, which are designed to prevent disease. According to Chinese medical tradition, the graceful

PINPOINTING THE PROBLEM This model ear shows the precise location of some three dozen acupuncture points. Each one is linked to a different part of the body.

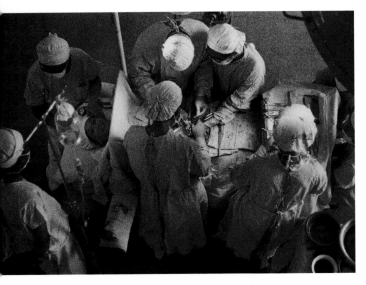

CONSCIOUS CUT Smiling for the camera, a Chinese patient undergoes an operation with two acupuncture needles in her ankle and knee providing the only anesthetic.

like the medical equivalent of a brush fire, becoming one of the most important forms of alternative medicine in the Western world today.

Since acupuncture became widespread in the 1970s, doctors have treated it with considerably more respect than homeopathy, and have devoted a considerable amount of research into finding out how it works. The reason for this difference in treatment lies in science. While homeopathy has no explainable scientific basis, acupuncture does involve physical changes that can be seen and measured. To western doctors, this

helped open a door into a new and remarkable area of medicine.

The experiments conducted since the 1970s show beyond doubt that acupuncture is effective in dealing with a range of problems, although in the West it has rarely been used in operative anesthesia. Western researchers have put forward a range of theories to explain how acupuncture may affect the nervous system, and how it may help the brain provide natural pain relief. However, as yet the significance of acupuncture points and meridians is still a mystery. Some seem to follow the alignment of nerves, but others do not seem to have any link with known physical structures. All that

SURGERY WITHOUT PAIN

In September 1971, during a period of rapidly thawing relations between the United States and China, E. Grey Dimond, an American doctor, was invited to see Chinese medicine firsthand. Dimond had a special interest in acupuncture. Writing in the *Journal of the American Medical Association*, he reported the remarkable effects of acupuncture when used as an anesthetic. In this operation—one of several that Dimond witnessed at Kwangtung Hospital in Kwangchow (formerly Canton)—the patient was a man who had a suspected cancer of the thyroid gland.

"There were no preoperative medications. The patient walked into the operating room, took off his pajama top, retaining the pants, and stretched out on the operating table. One stainless steel acupuncture needle was inserted in the extensor aspect [upper surface] of each forearm…A small clip was attached to the shaft of each needle and then connection made to a direct current battery power unit delivering 9 volts at 105 cycles per minute. Details of the wave form, current or circuitry could not be supplied by the anesthetist. An intravenous drip of 5 percent dextrose was begun and to it was added 50 mg of meperidine hydrochloride [a mild painkiller]. Typing and crossmatching had been done. During a 20-minute 'induction period' surgical preparation and draping were done. No other anesthetic agent was used. The patient remained fully conscious and normally alert. He advised me, through the interpreter, that he was noting numbness and tingling of both hands; no motor change [change in the ability to move] occurred. After 20 minutes surgery began and a skillful team moved rapidly through the operating procedure. At one point the patient took a sip of water. A large adenoma [glandular tumor], approximately 2 centimeters by 3 centimeters [$^3/_4$ x $1^1/_4$ inches] in size, was removed and the wound closed. The patient sat up, had a glass of milk, held up his little red book, and said in a firm voice 'Long live Chairman Mao and welcome American doctors.' He then put on his pajama top, stepped to the floor, and walked out of the operating room."

ALEXANDER TECHNIQUE
A therapist adjusts the curvature of a cellist's spine, using methods developed by Frederick Alexander.

scientists can say is that acupuncture is supported by "empirical evidence"—in other words, it has a measurable effect, but the mechanism behind it is unclear.

Posture and manipulation

Eastern medicine places great emphasis on the restoration of bodily balance and harmony. In the West, some more recent forms of treatment seek the same ends by making corrective adjustments to the patient's posture.

One form of alternative medicine was developed not by a doctor, but an actor. Frederick Matthias Alexander, born in Tasmania in 1869, began to lose his voice on

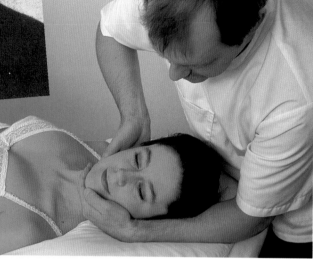

SWEEPING CLAIMS Andrew Taylor Still (left) believed that osteopathy could cure infectious diseases. Today, most osteopaths restrict their treatment to muscular and skeletal disorders.

HEALTH AND CULTURE

Cultural expectations and experiences play an important part in people's perceptions of health. For example, until the 1970s, French doctors often associated headaches and similar problems with *crise de foie*—literally, liver crisis—a condition that has no precise equivalent outside France. German doctors show a particular interest in blood pressure, while British patients often have an unusual preoccupation with the state of their bowels. As a nation, Americans show the greatest tendency to resort to surgery and other forms of highly interventionist treatment.

stage and sought help to overcome the problem. Orthodox medicine was unable to help, so Alexander tried to determine the cause of the problem himself. By observing himself in a mirror, he noticed that he automatically lowered his head when he tried to project his voice, tensing his neck and depressing his vocal cords. He decided that the way to overcome this was to re-educate his body by consciously altering his posture. Once he had learned his new pattern of behavior, his voice returned.

PUSHED INTO PLACE Using a technique called the rotational thrust, an osteopath applies pressure to a patient's spine to correct a disorder in the neck.

In 1904, Alexander emigrated to Britain and began to teach his methods to the acting community in London. His pupils included many famous theatrical names of the day, such as Henry Irving and Lillie Langtry, and he also attracted literary figures such as Aldous Huxley and George Bernard Shaw. When Alexander began teaching in New York, he found an influential supporter in Professor John Dewey, the educational philosopher, who did much to promote his work. What began as enlightened self-help grew into a completely new form of physical therapy: the Alexander technique. From its humble beginnings, the Alexander technique has gained wide acceptance as a form of alternative medicine, and countless patients, or "pupils," have learned to become aware of their posture, and to correct any distortions that have developed.

Like the Alexander technique, two other forms of posture-related medicine—osteopathy and chiropractic—were "invented" by individuals during the latter years of the 19th century, but found increasing favor in the 20th. Osteopathy was the creation of Andrew Taylor Still, who died in 1912. The son of a Methodist minister in Virginia, Still became a qualified doctor, but was devastated by the loss of three of his children during an epidemic of meningitis in 1874. He developed the theory that most diseases and disorders are caused by displacement of the bones or spine from their correct positions, and that careful manipulation of the body could restore it to normality and health. At the turn of the century, Still's ideas were refined by J.M. Littlejohn, a physiologist, who set up the British School of Osteopathy in 1917—a sign of the growing professionalization of alternative medicine.

Chiropractic was also developed in the

SIGNIFICANT SWING In pendulum therapy, movement of the pendulum is said to reveal the nature and location of ailments.

United States by D.D. Palmer, himself a practitioner of osteopathy. He claimed that almost all diseases are caused by neural malfunction and could be cured by vigorous manipulation of the spine.

Since the early 1900s, the relationship between osteopathy and orthodox medicine has developed by stages from suspicion to mutual respect. Osteopaths have absorbed

ESSENTIAL OILS In aromatherapy, selected plant oils are used to treat conditions such as muscular aches and depression.

many of the basic principles of orthodox medicine, and have expanded into areas of expertise that most doctors lack.

On the fringes of the fringe

Since Still's time, many more forms of treatment have been added to the list of alternative therapies. Some, such as rolfing, which was developed in California by Ida Rolf, are therapies that help the body by gentle massage and manipulation. Others, including autogenic training, or training in relaxation techniques, and biofeedback, in which the patient seeks to regulate involuntary functions such as heartbeat, are techniques that cultivate conscious control of the body, helping it to overcome stress. Autogenic training was developed in Germany in the 1930s, while biofeedback is more recent. It uses instruments such as pulse recorders or electrical-resistance meters to monitor the body. With their help, a person using biofeedback can learn to adjust their pulse or blood pressure in a way that avoids the need for drugs.

Neither autogenic training nor biofeedback are alternative systems of medicine, but they are forms of alternative therapy that many doctors

DISTANT HEALING Radionics is one of the few forms of therapy that claims to be able to act without the patient being physically present.

around the world now recognize as valid. The same cannot be said of some other products of the disaffection with 20th-century scientific medicine. Iridology, for example, is a diagnostic technique that asserts that all bodily disorders are reflected in the distribution of color in the eye, an idea that most orthodox doctors would dismiss out of hand. Radionics is more esoteric still—a system of diagnosis that works at a distance, with the practitioner using a pendulum or electronic instrument to "tune into" the patient.

As with iridology, no persuasive scientific evidence has yet been advanced that radionics has any reliable effect. But at the start of a new century, the interest that such therapies generate is often as revealing as the therapies themselves. It shows that, for many people, scientific medicine is not necessarily the only route to good health.

HEALTH AND HUMAN LIFE

BIRTH, GROWTH AND AGING ARE AS MUCH A PART OF HUMAN LIFE TODAY AS THEY WERE WHEN THE 20TH CENTURY BEGAN. WITH THE DEVELOPMENT OF MODERN MEDICINE, MANY OF THE DANGERS FORMERLY ASSOCIATED WITH BIRTH AND INFANCY HAVE BEEN OVERCOME. BUT LATER, DESPITE THE MANY ADVANCES THAT HAVE BEEN MADE IN MAINTAINING QUALITY OF LIFE, MATURITY AND OLD AGE STILL BRING WITH THEM MEDICAL PROBLEMS THAT SCIENCE HAS YET TO RESOLVE.

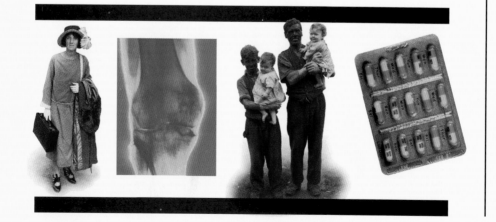

THE MIRACLE OF BIRTH

DEVELOPMENTS IN REPRODUCTIVE MEDICINE MEAN THAT HAVING A BABY IS MUCH SAFER AND—FOR MOST PEOPLE—MORE WITHIN THEIR CONTROL

In the opening decade of the 20th century, having a baby was a hazardous business. There were no antibiotics to overcome infections, and no blood transfusions in case of hemorrhage during birth. Few expectant mothers had their blood pressure checked, which meant that there was no early warning of toxemia, a disorder that can be deadly. Out of every 1,000 births in Europe and America, roughly five proved fatal for the mother, and in more than 150 cases the child died. At a time before the widespread use of contraception, women became pregnant much more frequently than today. Each pregnancy was like an entry in a lottery, with an accumulating chance that things would eventually go wrong.

Until the 1900s, medical involvement in childbirth was minimal. Some doctors played no part in it at all, considering it to be an unsuitable activity for persons of professional standing. Mothers from wealthy backgrounds were likely to have a doctor in attendance at the moment of birth, but poorer women were assisted by midwives, many of whom had no formal qualifications. Problems often developed because there was no system of monitoring the expectant mother's health as her pregnancy developed. The work of the doctor or midwife began at the moment labor started, and ended as soon as the child was born.

In 1901 J.W. Ballantyne, a Scottish obstetrician, took the first steps toward establishing a system of prenatal care. He examined expectant mothers several weeks before birth was due, and admitted them to the hospital in advance of birth if complications looked likely to arise. It was a novel approach, condemned by some doctors as unnecessary. But Ballantyne's initiative had a noticeable effect on the survival rates of his patients. Within a decade, the idea of prenatal care spread to the United States, and it gradually became a recognized part of medicine.

Before World War I, careful attention to hygiene helped to reduce some of the infections associated with pregnancy, but major improvements in this area did not come until the discovery of sulphonamide drugs in the 1930s and antibiotics a decade later. But as the century opened, developments were under way in another aspect of childbirth— the management of pain.

Twilight sleep

The use of anesthesia in childbirth first gained widespread acceptance in 1853, when Queen Victoria requested chloroform during the birth of her eighth child.

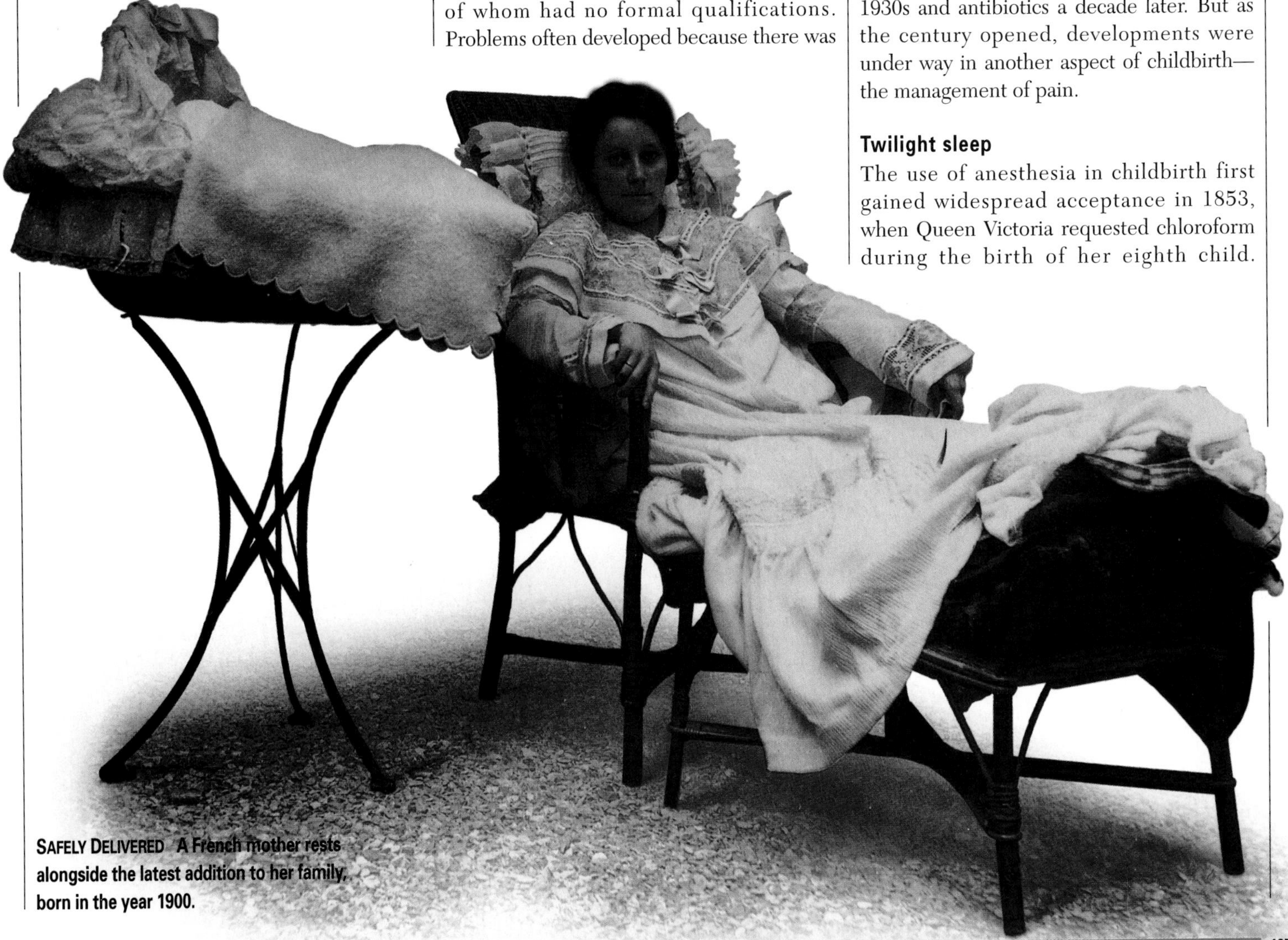

SAFELY DELIVERED A French mother rests alongside the latest addition to her family, born in the year 1900.

1900

1902 Twilight sleep used to lessen pain during childbirth

1909 Invention of first intrauterine devices (IUDs)

1916 Margaret Sanger opens first birth-control clinic in United States

1933 Publication of Grantly Dick-Read's Natural Childbirth

1950

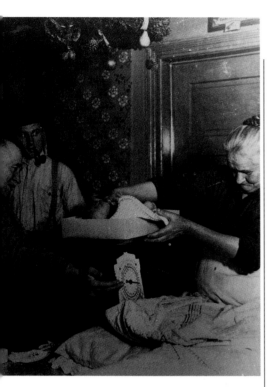

WEIGHING IN Before postnatal care became widespread, a baby's weight gave the most useful clue to its state of health.

However, 50 years later, anesthesia during labor was still problematic. Too much anesthetic could complicate the process of birth, and could even trigger a heart attack in some mothers.

In 1902, a group of German obstetricians devised a new way of dulling childbirth pain, called *Dämmerschlaf*, or "twilight sleep." The idea behind twilight sleep was not to abolish pain altogether: instead, it banished the memory of pain, so that the mother emerged from the process of childbirth without any of the painful associations that it could otherwise bring.

One form of the twilight-sleep technique, developed at a clinic in Freiburg, involved two different drugs—morphine, which dulls the perception of pain, and scopolamine, which erases short-term memory. Once the process of labor began, the expectant mother would be given morphine, and then a series of injections of scopolamine, with memory tests after each one. If the treatment was carried out correctly, the patient became drowsy but did not lose consciousness. The mother was placed in a darkened

SHARED EXPERIENCE In the 1950s, improvements in analgesia meant that more mothers could remain conscious to experience the moment of birth.

room and the medical staff around her took care to make as little noise as possible—an approach designed to remove any cues that might trigger memories once birth was over.

Twilight sleep soon found enthusiastic supporters on both sides of the Atlantic, particularly among the upper classes. In 1909, a Mrs. C. Temple Emmet of New York, who was connected to the wealthy Astor family, became the first American woman to undergo a "twilight" delivery at the Freiburg clinic. She was delighted with the results, and on her return to the United States became instrumental in setting up the National Twilight Sleep Association, which pressed for the technique to be used in America. Another advocate, Hanna Rion Ver Beck, wrote that upper-class women were particularly prone to nervous shock in childbirth—an idea that was widely held at the time. Twilight sleep offered them a way out of their suffering.

Many doctors, in both the United States and Europe, had serious reservations about the technique. When carried out with scrupulous attention, as in the Freiburg clinic, twilight sleep did seem to produce good results, but if drug doses were even slightly miscalculated the outcome could be quite different. Instead of being relaxed, the mother could become restive and difficult to manage, so that every minute of labor

required constant vigilance. It was far more than most doctors could cope with, and despite vigorous campaigning by women's groups, twilight sleep fell into disuse.

While the technique ultimately failed, the debate surrounding twilight sleep proved to

APGAR SCORES

In 1953, Virginia Apgar, an American anesthetist, introduced a simple scoring system that allows doctors to establish the condition of babies immediately after birth. The system is based on the baby's color, its breathing and crying, muscle tone and heart rate, and also its reaction to being touched on the sole of the foot. A score of between 7 and 10 indicates a healthy baby.

be a landmark. For the first time, women had dared to take on the medical profession and demand a say in the way they were treated. It was a theme that would recur as the century progressed.

Timetable births

With the failure of twilight sleep, doctors placed increasing reliance on drugs that abolished pain during childbirth, rather than ones that made mothers oblivious to it. By degrees, childbirth became an increasingly medical matter. In the 1920s, only one in five births in Britain took place in hospitals. By the early 1950s, the figure rose to nearly

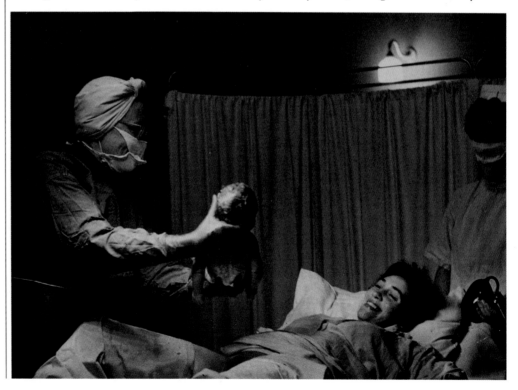

| 1950s Drugs used to induce birth | 1960 Contraceptive pill goes on sale in the United States | 1960s Ultrasound scanning of fetuses becomes routine | 1969 Human cells are fertilized outside the body | 1978 Birth of Louise Brown, the first "test-tube" baby |

forceps, which are placed around its emerging head.

In this climate of intervention, obstetricians also began to look at the timing of birth. In cases where complications threaten—for example, if the mother has a narrow pelvis—birth is sometimes safer if it occurs before the normal gestation period is complete. For centuries, midwives artificially triggered labor by using ergot, a fungus

DELIVERY FORCEPS
Instruments like this were once routinely used, clamped around a baby's head, to speed the process of birth.

that grows on rye. This triggering process—called induction—used to be employed only in genuine emergencies, but in the 20th-century hospital environment its use gradually became routine.

Ergot remained one of the most important induction agents, and its derivatives are still in use today. However, over the last 75 years, several other chemical agents have been employed to initiate labor. They include quinine, which was used until the 1930s, and oxytocin, a hormone produced by the human pituitary gland.

Oxytocin was identified in 1906, and was first synthesized artificially in 1953. It is normally released in large amounts just before

NEW IN TOWN Newborn babies lie in their cribs at New York's Beth Israel Hospital in the 1950s.

birth, making the muscles of the uterus contract. These contractions trigger the release of more oxytocin, and this system of positive feedback continues and builds until birth is complete. By giving expectant mothers synthetic oxytocin, obstetricians in the 1950s became able to induce birth at a specific time. It could be a time that was appropriate for medical reasons; equally well, it might be one that fitted more conveniently into the hospital schedule.

Once synthetic oxytocin became available, it opened the way for even greater management of the process of birth. The hormone not only proved useful for triggering labor, it could also be used like an accelerator pedal, to adjust the time labor lasted. During the 1950s and 1960s, research showed that mortality was lowest when the first stage of labor lasted between 12 and 24 hours in women who were having their first child, and between 3 and 24 hours in women who already had children. Armed with this knowledge, obstetricians embarked on a process of active management, which ensured that all births followed a fixed timetable. Any mother who showed signs of falling behind was given extra oxytocin to keep her labor on course.

Surgery and birth

In the 1950s, the rise of active management was accompanied by other changes in hospital delivery rooms. One was the invention of the vacuum extractor, or ventouse. Despite its ominous-sounding name, this instrument was devised as a safer alternative to obstetric forceps, for use in cases of difficult delivery.

two-thirds, a change mirrored in all other developed countries.

During this process of "medicalization," doctors and obstetricians took greater charge over processes that previously had been left to follow a natural course. In the 1920s Joseph B. DeLee, an influential American obstetrician from Chicago, advocated complete anesthesia as soon as labor began, followed by manual delivery of the baby using

WE DIDN'T KNOW NOTHING

Ignorance about sexual matters was widespread at the beginning of the century, and despite progress in education, lack of knowledge in this area proved hard to eradicate. During the 1960s, a group of doctors in Louisiana set up the first state-wide family planning program, and interviewed hundreds of women about reproductive matters. These are some of the replies they received when they asked when and how a woman was most likely to get pregnant:

"It all depends on how long you have been without a sex[ual] relation[ship]. If you are away from a man and then go back to having sex relations again you get pregnant straight away." (Age 25)

"I think anytime you sleep with a breathing man you've had it." (Age 37)

"There is a certain time of the month. I don't know how to count it. It looks to me like you feel more loving at certain times of the month. This is the time you can get pregnant." (Age 37)

"Because it just happens with the nature of God. If He wants to bless us, we shouldn't complain." (Age 34)

"Anytime, I know that eggs comes down and stays there just waiting to be fertilized." (Age 28)

"One of the freaks of nature, I guess. It is nature itself. If nature want you to have it, you will." (Age 26)

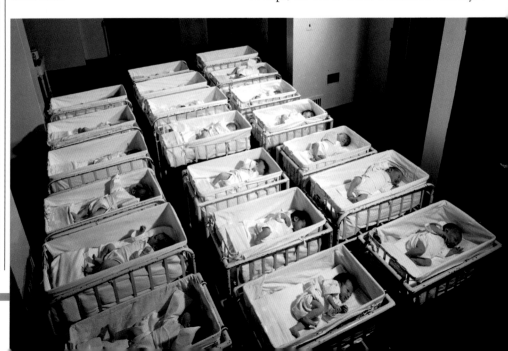

Forceps clamp around the baby's head, and can cause damage to the mother's birth canal. The ventouse, on the other hand, has a suction cap that clings to the baby's scalp when a vacuum pump is switched on. The cap is then used to draw the baby slowly out of its mother with much less risk of damage.

With the advent of ultrasound imaging in the 1960s, doctors and obstetricians found it

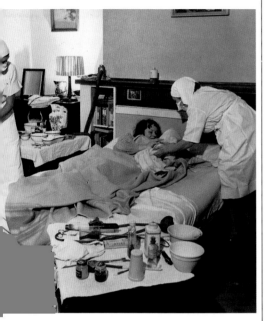

BIRTH AT HOME In the 1950s, scenes like this became increasingly rare as women were persuaded to give birth in the hospital.

much easier to identify problems such as breech presentation, which can cause serious difficulties during birth. In a normal pregnancy, the child often develops in a head-up position, but eventually turns itself upside down so that its head emerges first. In a breech presentation, the child remains head-up and emerges bottom-first, sometimes getting stuck on the way. If this does happen, the age-old remedy is to perform a Caesarean section, an operation that delivers the baby through an incision cut in the mother's abdomen.

At one time, Caesarean section was used only as a last resort, when the lives of both mother and baby were in grave danger. But with the development of sterile surgical procedures, and the discovery of antibiotics several decades later, Caesarean section became a safe alternative to normal birth.

The effect of this change was remarkable. At the beginning of the century less than 2 percent of deliveries were carried out by

SURVIVING EARLY BIRTH

Babies are normally born after a gestation period that lasts about 40 weeks—roughly nine months. Premature babies are those born before 37 weeks have elapsed, at a time when their physical development is not yet complete. At the beginning of the century, most premature babies had only a slim chance of survival. Today, thanks to improvements in neonatal medicine and technology, nine out of ten babies born as early as the 30 week stage succeed in their battle for life.

One of the most important steps toward saving premature babies was taken in the early 1900s, with the development of the incubator. Because they have such small bodies, premature babies face serious difficulties in controlling their body temperature. An incubator does this for them, by maintaining

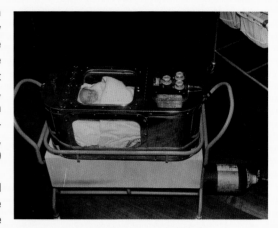

TEMPORARY HOME This metal-cased incubator, designed in 1946, had glass windows, electric heating and an on-board oxygen supply.

a constantly warm atmosphere. Before the First World War, early incubators worked with the help of water-filled jackets. Later, electrically heated incubators, forerunners of incubators used today, were introduced.

Premature babies often have difficulty breathing because their lungs lack a wetting agent called surfactant, which normally lines the inner surface of each lung and prevents the lung's microscopic air sacs from sticking together when air is expelled. This lack of surfactant often means that the air sacs remain partly stuck together and cannot fully inflate, leading to a condition called respiratory distress syndrome, or RDS.

Beginning in the early 1940s, RDS was treated by flushing the air in incubators with a high level of oxygen. However, within a few months serious side effects began to appear. Dr. Theodore Terry, an eye specialist in Boston, examined several premature babies who had developed an unusual form of blindness, caused by fibrous tissue forming within the eye. By 1945 he had seen more than 100 cases of this previously rare disorder without knowing the cause. In the years that followed, retrolental fibroplasia, as the condition was called, was found in babies across the world. In 1951, evidence appeared that blindness was rare in premature babies

AIDING YOUNG LUNGS Born three months early, this infant will stay in an incubator until its lungs have fully developed. A respirator keeps it alive.

given less oxygen, and a medical study eventually proved the connection. By the mid-1950s, oxygen levels were reduced, and retrolental fibroplasia all but disappeared.

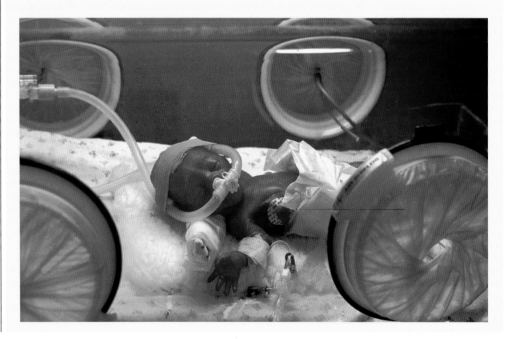

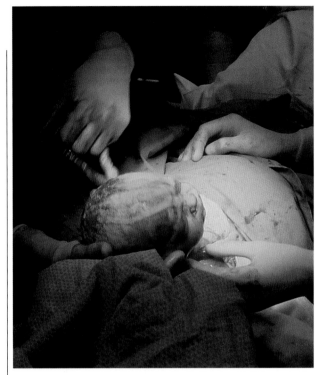

CAESAREAN BIRTH Surgical intervention in birth has increased dramatically in the last 40 years, prompting a debate about how far it is justified.

Caesarean section, and they often proved fatal to the mother. Ninety years later, the proportion in many European countries had risen to over 10 percent, while in North America, it grew to almost epidemic proportions, accounting for up to 25 percent of births. Many of these operations are performed on valid medical grounds, but in the United States particularly, critics of the steep rise in childbirth surgery point to another explanation behind the changing numbers. By advising Caesarean sections, these critics say, American doctors avoid many of the potential problems associated with difficult deliveries—and the danger of being sued if things go wrong.

The natural childbirth movement

From its early days in the 1920s and 1930s, interventionist treatment came in for some harsh criticism from both inside and outside the medical profession. One of the most influential of these critics was a British obstetrician, Grantly Dick-Read. In 1933, Read published *Natural Childbirth*, a book that was to have a profound effect on readers for decades to come.

In stark contrast to the prevailing trend of the time, Dick-Read protested what he saw as the excessive use of drugs and surgery during childbirth. "Nothing is to be more abhorred than deep anesthesia—forceps deliveries of normal babies, blue and flabby babies who will not cry, babies drugged and babies anesthetized. These pictures, so common in modern practice, are deplorable blunders of both judgment and action."

In the place of routine intervention, he focused on the positive role that relaxation could play in breaking the link between fear, tension and pain. This kind of relaxation, he explained, was quite different to the knockout effects of a powerful anesthetic, which rob a mother of consciousness without affecting subconscious tension. "One important result of practiced relaxation is the complete obliteration of mental imagery. Without the vision of that which terrifies, peace may easily be acquired. During labor, peace is synonymous with happiness, and the happiness of labor is the perfect completion of nature's most delightful function." To help bring about this relaxation, Dick-Read recommended specific exercises that could be carried out in the weeks leading up to birth.

Dick-Read's book was received with skepticism by many members of the medical profession, even though he had been careful not to rule out interventionist treatment when it was clearly necessary. However, among women, it clearly touched a chord. By the time he published a second book on the subject in 1942, *Childbirth Without Fear*, the idea of pain control through relaxation had gained a strong following among mothers-to-be. The reaction was even greater than that which had surrounded twilight sleep, but this time women were arguing against the use of drugs, rather than for them.

After a sluggish start, Grantly Dick-Read's books eventually became best-sellers, and in the decades that followed, they led to a wave of developments in "natural" birth methods that involved a minimum of medical intervention. Among them were techniques advocated by the French doctor, Frederique Leboyer, who wrote *Birth Without Violence* in the 1960s. Like several other professionals in his field, Leboyer believed that the moment of birth had a profound impact on the future development of a child, making it essential that it was as gentle as possible. He recommended that birth should take place in tranquil, dimly lit surroundings, but he also suggested that it could take place in water, an idea that quickly grabbed the headlines.

During the closing decades of the century, the initial antagonism between the followers of natural and "medical" childbirth has gradually given way, as each side has come to see some of the merits in the other's case. The result today is that many women use natural relaxation techniques to prepare for birth, knowing that medically based pain relief is available should they need it.

Birth control

In the final years of the 19th century, the average woman in rural parts of North America or Europe could expect to become pregnant some nine or ten times during the course of her reproductive life. Although contraceptives existed, they were illegal in many countries and the only reliable method of limiting family size consisted of sexual abstinence. Pregnant women desperate to

BORN IN WATER A mother holds her new baby, born in a birthing pool—a method of delivery that first became popular in the 1960s.

avoid having further children often resorted to illicit abortions, sometimes with calamitous results.

In 1914 Margaret Sanger, an American nurse, decided to address the problem of large family size and the poverty and health problems that it often brought. In her book *Family Limitation*, Sanger advocated the idea of birth control—a term she herself coined—and discussed contraception in everyday terms. Official reaction was hostile, and the Federal government tried to prosecute her for obscenity. But Sanger was unrepentant. In 1916 she opened her first clinic in Brooklyn, modeled on similar clinics she had seen in Holland, a move that earned her a sentence of 30 days in a city workhouse. It also proved to be the beginning of a movement that was to sweep the United States,

SIGN OF THE TIMES In America, as in other industrialized countries, as more children survived and before the advent of birth control, a large family was not unusual.

and the rest of the developed world.

Sanger's activities were soon mirrored on the other side of the Atlantic, by a remarkable woman from a completely different background. Marie Stopes had trained as a botanist, and had a particular interest in fossil plants. At the age of 25, she became the youngest-ever Doctor of Science in Britain, and she looked set to make noteworthy progress through the relatively untroubled world of academia. But in 1911 she married, an event that changed her life.

Sexually, the marriage was a disaster, because her husband turned out to be impotent. In common with most women of the day, Stopes knew little about sexual matters, but with characteristic thoroughness she decided to educate herself, and to pass on the fruits of her research to other married couples. The outcome in 1918 was a book called *Married Love*.

Today, parts of *Married Love* make somewhat comical reading, because Stopes had

TEST-TUBE BABIES

CONCEIVED IN A TEST-TUBE RATHER THAN HER MOTHER'S BODY, LOUISE BROWN'S BIRTH HERALDED A REVOLUTION IN THE BATTLE AGAINST HUMAN INFERTILITY

On July 25, 1978, the world's first "test-tube baby" was born at Oldham and District General Hospital in England. At 5.7 pounds, Louise Brown weighed slightly less than average, but in all other respects she was a normal, healthy child.

Louise's birth was the culmination of many years' work by two British researchers, Patrick Steptoe and Robert Edwards, who eventually joined forces in 1968. While other scientists carried out research into the development of animal egg cells, Steptoe and Edwards investigated human egg cells. Their hope was that these cells could be fertilized outside the body, providing a treatment for some forms of infertility.

In normal circumstances, ripe egg cells develop at the rate of one every 28 days, and are released by a woman's ovaries. Each egg cell is swept toward the womb, or uterus, along a duct called a Fallopian tube. After sexual intercourse, sperm cells travel up the uterus and into the Fallopian tubes, where they meet the egg. If a sperm cell successfully fertilizes the egg, development begins.

To attempt fertilization outside the mother's body, Steptoe and Edwards first had to obtain egg cells at exactly the right stage of development. They did this by giving their patients hormones that increased the rate of egg-cell production. A small incision was then made in the patient's abdomen, and an endoscope inserted and guided toward one of the ovaries. An instrument called an aspirator was then used to suck up the egg cells, just when they were ready to be released.

As work progressed, Steptoe and Edwards became more skilled at collecting egg cells and keeping them alive, but their work was already attracting criticism. Some clerics and politicians called for all research on human egg cells to be abandoned, and when Steptoe and Edwards submitted a paper describing their aspirator to a leading medical journal, it was

STAR PLAYER Louise Brown crawls across a table during an American television broadcast in 1979.

turned down on ethical grounds. Despite a growing chorus of criticism, they carried on.

The next challenge was to fertilize the egg cells they had collected. In animal experiments, researchers had found that just mixing eggs and sperm cells would not work, because normally sperm cells have to undergo chemical changes inside a woman's body before they can fuse with an egg. This process, called capacitation, initially proved very difficult to bring about in the laboratory, and it took Steptoe and Edwards several years to find a successful method. This consisted of treating the sperm cells with a mixture of chemicals found within the female reproductive tract. By 1969, fertilization of human eggs had been observed

SMALL BEGINNINGS A human embryo at the 16-cell stage, seen against the tip of a pin. At about this stage, the embryo normally implants itself in the mother's uterus.

in a test tube. Some of the eggs went on to divide into several cells—the prelude to the development of an embryo.

In natural circumstances, a fertilized egg divides several times on its journey down the Fallopian tubes, and then implants itself in the lining of the uterus. Steptoe and Edwards had to mimic the first part of this process in the lab, keeping the egg cells alive for more than three days while the vital divisions took place. When each egg, now a microscopic embryo, reached the 16-cell stage, they could implant it in the mother. At this point the lab work was over and normal development could resume.

Since 1978, hundreds of children have been born by the technique that Steptoe and Edwards pioneered. For many couples unable to conceive naturally, it has swept aside an apparently insuperable obstacle to raising a family.

MORE TEST-TUBE TRIUMPHS Patrick Steptoe and Robert Edwards seen in 1985, shortly after the birth of a baby from a frozen embryo.

RELENTLESS CAMPAIGNER
Brushes with the law failed
to dampen Margaret
Sanger's enthusiasm for
promoting birth control.

such a dearth of first-hand knowledge about her subject. She refers to mysterious "vital energies" and "nerve-forces" that come into play during attraction between men and women, and describes how a man's sexual urges can be usefully diverted into "creative work" when his partner is not in a receptive frame of mind. But despite its mixture of wishful thinking and educated guesswork, *Married Love* could also be candid and direct. It discussed taboo subjects such as orgasm and masturbation, and made it clear that sex could be a pleasurable activity for both men and women. It sold over a million copies, and went through more than two dozen editions.

Like Margaret Sanger, who she met in 1915, Marie Stopes was not only a writer; she also became active in promoting family planning in a practical way. She opened her first clinic in London in 1921, and in 1923 wrote a manual on contraception designed specifically for the medical profession. In addition to describing the various methods of contraception available at the time, she also reviewed their legal position in different countries. In most European countries except Britain, contraceptives were still outlawed; in the United States, meanwhile, they were grouped with other kinds of "obscene or indecent matter." If they were sent by mail, the sender and recipient were both liable to a maximum of five years in jail, or a $5,000 fine.

After the 1920s, bans on contraception were slowly lifted in many western countries, although religious prohibitions remained, and still do, particularly for Roman Catholics. During this period, methods of contraception changed. Early "barrier" methods, such as condoms and diaphragms, which existed in the 19th century, were gradually improved so that they became more reliable, safer and easier to

use. Intrauterine devices, or IUDs, which were invented in 1909, underwent a proliferation of designs as researchers sought to perfect a shape that would prevent fertilization without damaging the lining of the uterus. In the 1930s, progress was also made in the "calendar" or "rhythm" method of birth control, which relies on women's estimates when they are least likely to conceive. But the greatest change of all came in 1960, with the launch of the oral contraceptive, or "pill."

The pill

Like the birth-inducing agent oxytocin, the pill is based on a hormone that plays a natural part in human reproduction. The hormone

MENTIONING THE UNMENTIONABLE Marie
Stopes's book *Married Love* made her one of
the best-selling writers of her time.

Diaphragm

Cervical cap

Saf-T-Coil

IUD

Condom

A CENTURY OF CONTRACEPTION
Methods such as the condom
and diaphragm have been joined
by IUDs (intrauterine devices).

concerned, called progesterone, is produced by a woman's ovaries. Through a complex chain of interactions, it ensures that ovulation (egg-cell release) occurs only once every 28 days, approximately. Progesterone also ensures that it shuts down during pregnancy.

Early in the century, biochemists discovered that hormones from a mammal's ovaries could prevent egg-cell release, but pure progesterone proved difficult to isolate. The German chemist Adolf Butenandt managed to prepare a tiny amount in 1934, but progesterone was still an extremely rare substance, making it tricky to investigate in the laboratory. Thus it might have remained for many years, had it not been for an extraordinary breakthrough in the 1940s. While he was working in the forests of southern Mexico, Russell E. Marker, an American biochemist, discovered that he could prepare a close derivative of progesterone from a species of wild yam. Suddenly, progesterone could be mass-produced. Within a year,

Marker had made more than 6 pounds—an unheard-of amount of any hormone, whatever its origin. At the prices current at the time, it was worth nearly $250,000.

Marker set up a company to manufacture

progesterone, but then lost interest in the project and abandoned the world of biochemistry altogether. His former associates continued the work, and in the early 1950s they successfully converted the original hormone into a form that was not inactivated by being swallowed. Samples were sent to the United States for testing, and it was soon established that they could prevent ovulation. However, at this stage, its use for con-

traception was still far from assured. Some of the research team had reservations about the ethics of hormonal contraception, and others were concerned about the possible side effects of healthy women taking hormone supplements for long periods of time.

The worries on the second score were largely settled in the mid 1950s by a clinical trial held in Puerto Rico—an American territory in which contraception was legal. By 1957 the "pill" had shown itself to be effective and safe and was approved in the United States as a treatment for menstrual irregularity. In 1960 Enovid, the first commercially available form of the pill, went on sale.

Initially, the commercial contraceptive pill was purer than the original hormone preparation; it contained progesterone but no other active ingredients. Ironically, this pure hormone was found to cause problems in some women, and turned

CRUCIAL KNOWLEDGE A doctor at a Jordanian clinic explains to a mother how to use the contraceptive pill, which comes in many forms (right).

out not to be entirely reliable. As a result, researchers returned to the original formulation, which contained traces of a second hormone, a form of estrogen. However, in a small proportion of women this formulation turned out to produce side effects of its own, including strokes. In the 1970s, pill manufacturers reduced the amount of estrogen, creating the lower-dose formulations that are used today.

From the moment it was launched, the pill has generated a phenomenal mixture of acclaim and condemnation. To its supporters, it has given women control over their own

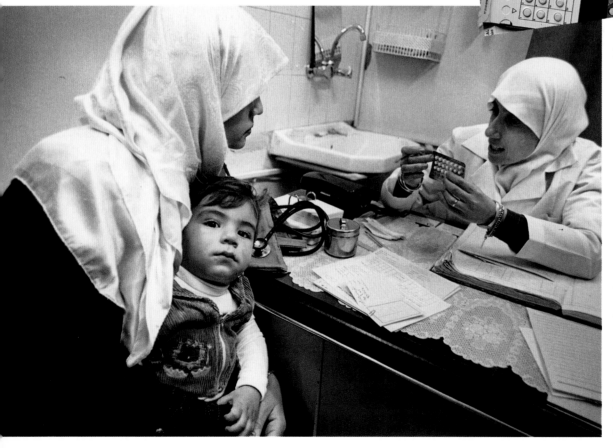

fertility, helped to reduce the poverty associated with large families, and played a part in restraining—if only by a small amount—the huge rate of growth of the human population. But in the eyes of its critics, the pill has been almost wholly harmful. By allowing sexual freedom it has loosened the strength of the family, and has created a society in which sex is seen more as a recreational activity than as an expression of the deep ties that exist between permanent partners.

Treatment for infertility

Paradoxically—considering all the work that has gone into devising better contraceptives—humans are surprisingly bad at conceiving compared to our close relatives in the animal world. For

THE MIRACLE OF BIRTH

THE POPE AND THE PILL

On July 29, 1968—eight years after the pill became commercially available—Pope Paul VI issued an encyclical called *Humanae Vitae* (Human Lives). In it, he proclaimed the Roman Catholic Church's rejection of all forms of contraception, with the exception of those such as the rhythm method, which rely exclusively on natural means. The encyclical led to immediate divisions in the Catholic clergy, with several bishops resigning. In the years since the ruling, it has continued to cause controversy, and although it remains an article of Roman Catholic teaching, the encyclical is ignored by many members of the Church.

CHURCH PROTESTS Supporters of the Pope gather outside London's Roman Catholic cathedral in 1968.

work. This was no easy matter, because sperm cells undergo subtle chemical changes as they make their way through the female reproductive tract. Without these changes, they cannot fertilize an egg.

By 1959, researchers had managed to bring about in vitro fertilization in rabbits, leading to the birth of healthy young—an achievement that attracted relatively little public attention at the time. In 1969, human eggs were fertilized outside the body for the first time. Nine years later, fertilization and reimplantation were successfully carried out on a human patient. The result was the birth of the world's first "test-tube" baby, Louise Brown, born in England in 1978.

Despite the fact that in vitro fertilization does not have a high success rate—in Britain in the late 1980s only about 10 percent of treatments were successful—it has become an accepted part of fertility treatment. Yet the ethical issues that surfaced in the late 1970s have not gone away. National laws have struggled to keep up with matters such as the ownership of eggs and embryos, and the legal rights of surrogate mothers, who undergo pregnancy on other women's behalf. Doctors find themselves addressing complex ethical questions, particularly when they help both to create life by tackling infertility and to end it by authorizing abortions. For many people, this dilemma is one of the most perplexing in modern medicine.

some couples conception is not merely difficult, it is practically impossible, creating the frustration and anguish of infertility.

In 1910, the first scientific attempt was made to investigate infertility with the invention of an X-ray procedure called hysterosalpingography, a tongue-twisting term derived from the Greek words for the womb and Fallopian tubes, which connect the womb with the ovaries. To produce the X-ray picture, an opaque dye is introduced into the woman's reproductive tract, so that any structural problems are revealed. In the 1930s, research also began to focus on male involvement in infertility—a break with the centuries-old tradition that the problem always lay with the "weaker sex." For the first time, analysis of the shape and numbers of sperm cells provided guidelines that helped to identify infertile men.

However, even by the 1960s, with the world's first heart transplant on the horizon, remarkably little was known about the early stages in the process of conception. Sperm cells are produced in such profusion that it is easy to examine them, but egg cells are a different matter. In humans, they are released by the ovaries at the rate of just one a month, and although they are much bigger than sperm cells, they are still microscopic and immediately vanish into the relative vastness of the female reproductive tract, on a journey along the Fallopian tubes to the womb. They only become detectable after successful fertilization when, several weeks

later, a growing embryo begins to take shape.

In medicine, ideas are often far in advance of practical methods. This was the case for the most revolutionary development in the struggle to overcome childlessness—

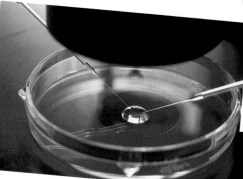

IN VITRO FERTILIZATION Sperm are collected from a donor sample (above). A sperm fertilizes an egg cell (right).

fertilization of egg cells outside the mother's body. This technique, called in vitro fertilization or IVF (fertilization in glass), was first suggested in 1937, but was not brought about in animals until two decades later. Obtaining the microscopic egg cells was only one of several major difficulties that lay in its path. Once the eggs had been extracted, they had to be kept alive and maintained in exactly the right kind of environment for fertilization to

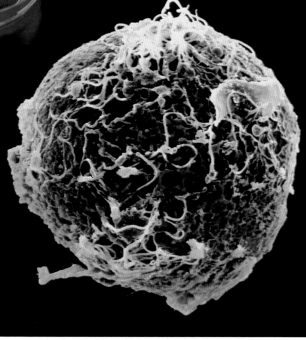

ATTACK FROM WITHIN

DESPITE SUCCESSES IN OTHER FIELDS, MEDICINE HAS YET TO CONQUER CANCER—A DISEASE IN WHICH THE BODY'S ENEMY IS ITSELF

During the first half of the century, cancer was an almost unmentionable disease. The word rarely cropped up outside medical journals, and newspaper reports went to great lengths to avoid any explicit reference to it. Public figures were never described as dying of cancer; instead obituaries spoke of "prolonged" illnesses, without actually stating what the illness was. There was a conspiracy of silence, fueled by a mixture of powerlessness and fear.

Nowadays, cancer is openly discussed, and there is no longer any sense of shame attached to suffering from the disease. Much more is known about the different types of cancer, and the steps that can be taken to prevent some of them from taking hold. But despite the vast amount of time and money devoted to research, fail-safe cures are not yet in sight. The mortality rates for some kinds of cancer have dropped quite sharply, but in developed countries up to one in three people die from the disease—a higher figure than when the century began.

One reason for this rise is that cancer, with some notable exceptions, is usually a disease of later life. With more people surviving to old age, more people run the risk of developing cancer eventually. Another reason lies with changes in the way we live. As researchers discovered in the first decades of the century, cancer is linked with substances in our surroundings—substances that are a side effect of life in an industrialized world.

Making sense of cancer

Cancer gets its name from the Latin word for a crab—a reference to the crab-like pattern of veins that sometimes forms in cancer of the breast.

EARLY TREATMENT A cancer patient is given an injection of serum in 1910. It probably did little to help.

Until the 19th century, cancer was thought to result from imbalances in the body's fluids, or from inflammation, and it was only when scientists began the detailed study of cells in the 19th century that its true nature was established. By 1900, pathologists knew that cancer is not a single disease, but encompasses many. The one feature that these diseases share is that they involve rapidly multiplying cells—cells that have somehow side-stepped the body's normal

CAPTIVE ALLIES From early in the century, laboratory rats and mice have played a key part in cancer research.

controls and grow in an unrestricted way. Having dismissed the idea that inflammation triggers cancer, turn-of-the-century medical researchers tried to unearth a more plausible cause. They began their work at a time when bacteria had been implicated in many diseases, and when new kinds of infectious agents—viruses—were being linked with others. With these developments still fresh, it seemed quite likely that cancer, too, might be caused by some kind of infection.

In 1911, this theory was given an early boost when Francis Peyton Rous, an American doctor, published the results of tests he had carried out on chickens. While he was working at the Rockefeller Institute in New York, Rous examined a chicken with a tumor. When the bird died, he

1906 World's first "radium institute" planned in Paris

1911 Rous concludes that cancer is caused by a virus

1915 Japanese researchers conclude that coal tar causes cancer

1931 Britain introduces regulations governing asbestos production

liquidized some of the tumor tissue and passed this through an extremely fine filter to exclude any tumor cells or bacteria. When he scratched the fluid into the skin of healthy chickens, they also developed cancer. Rous concluded that a virus, which is a non-cellular organism, was making some of the chickens' cells cancerous.

Rous's work proved to be well ahead of its time, and its importance was not recognized until more than half a century later when, at the age of 85, he was awarded a Nobel prize. During the last 20 years, researchers have shown that a number of cancers are caused by viruses and other microorganisms, and it now seems likely that infections cause at least 10 percent of cancer deaths. Rous's work initially led to a dead end, however. Cancerous cells themselves cannot pass on the disease to healthy cells, so attempts to produce immunity to cancer failed. As a result, the idea of infection as a cause fell out of favor, taking many decades to resurface.

With infection more or less ruled out, researchers looked for other external factors that might trigger cancer. In 1761, an English doctor had noticed that nasal cancer seemed to be linked with snuff-taking, and in 1775 his compatriot Sir Percivall Pott suggested that soot was the cause of scrotal cancer in chimney sweeps, who in those days often climbed up inside chimneys to do their work. By 1900, snuff and soot were not the only substances under suspicion. Industrial processes had created hundreds of new chemical compounds, and growing health problems, such as bladder cancer in synthetic dye workers, strongly suggested that some of these compounds were carcinogenic.

The first scientific demonstration of this effect came in 1915, when two Japanese researchers, Katsusaburo Yamagiwa and Koichi Ichikawa, investigated coal tar—a black, sticky liquid produced when coal is heated in ovens sealed off from the air. Yamagiwa and Ichikawa repeatedly painted coal tar on the ears of rabbits, and found that cancers often formed as a result.

Coal tar was—and still is—an important industrial raw material, but instead of being a single chemical substance, it is a mixture of many. The next step in the carcinogen hunt

was to isolate individual chemicals and test their effects. In 1930, after several years of painstaking work, researchers identified dibenzanthracene as the first pure chemical compound known to cause cancer. Dibenzanthracene is not actually found in coal tar itself, but a similar compound, called benzopyrene, which three years later was found to be carcinogenic, is one of the ingredients of coal tar. Since the 1930s, coal tar—and many other substances used in industry—have yielded a clutch of potent carcinogens. So, too, have some of the substances that we

PROFESSIONAL HAZARDS Twentieth-century medical research identified carcinogens in soot, smoke, coal tar and dozens of industrial processes.

use in daily life. At first glance, coal and tobacco may not seem to have much in common, but both are built of carbon-containing compounds; more importantly, both produce toxic benzopyrene when they are incompletely burned. It is therefore all the more ironic that while chemists were worrying about the effects of coal tar, a potentially far more dangerous source of carcinogens—the cigarette—was being smoked in ever-increasing numbers.

In 1900 cigarette smoking was relatively unpopular. In the United States, the average annual consumption of cigarettes was less than ten packs a year. By the early 1960s, however, it had leapt to over 200. As far back as 1912, some doctors had suggested that cigarettes might cause lung cancer, but lacked firm evidence for a link. There was no shortage of evidence about the rising incidence of lung cancer itself. Between 1900 and the mid-1950s, the death rate from lung cancer in

CANCER AND DIET

During the 20th century, research revealed that several forms of cancer are linked to the things people eat. In Japan, for example, the traditional national diet is low in fat, and so are the rates of breast and colon cancer. In America, the average diet is much higher in fat, and these two kinds of cancer are much more common. The children of Japanese immigrants in America who eat a Western diet have the same incidence of breast and colon cancer as other Americans.

1953 Watson and Crick unlock chemical structure of DNA

1971 National Cancer Act

1978 Weinberg triggers cancer in mice by transferring single genes from one animal to another

American men increased more than 50-fold, transforming this form of cancer from a relatively rare disease into a major killer.

The first large-scale studies linking lung cancer to smoking appeared in the early 1950s. In the early 1960s, the U.S. Surgeon General and the Royal College of Physicians in Britain officially endorsed the connection, and within a decade, cigarette sales in many countries began to fall. At the dawn of the 21st century, the lung cancer epidemic looks set to abate in most developed countries, but with changing patterns of cigarette consumption, it may well move on to other parts of the world.

Chromosomes and cancer

When carcinogens were first identified, no one had any idea how they acted. The same was true for cancers caused by viruses, and

HAZARDOUS INHERITANCE Early in the century, some biologists guessed that inherited genes might play a part in cancer.

CHILD'S VIEW This award-winning 1990 parody of a cigarette advertisement was drawn by a 12-year-old New Yorker. Below: X-rayed lungs reveal the signs of disease—an oval shaped carcinoma (orange-colored area) in the left lung.

also for the most common cancers of all— the ones that seem to appear spontaneously, without any prompting from outside.

In 1914, just a year before his death, the German cell biologist Theodor Boveri put forward the idea that cancer is triggered by abnormalities in chromosomes, the threads of genetic material that are found in the nuclei of all living cells. It was accepted that chromosomes functioned as a kind of biological instruction bank. From this, Boveri correctly deduced that any abnormalities—or mutations—in chromosomes might have far-reaching effects on the way in which cells grow, triggering some to become cancerous.

This theory did away with the idea that cancer might have a single cause. Boveri thought, correctly, that cancer-producing mutations might crop up in a wide variety of ways. Some could be the result of contact with external factors, such as carcinogens. Others might be present from the moment of conception, having been passed on from parents to children.

In the years following Boveri's death, chromosomes came under increasing scrutiny, and a growing body of evidence began to lend weight to the ideas that he had put forward. By the 1930s researchers had discovered that chromosomes consisted of proteins, together with a substance called deoxyribonucleic acid, or DNA. The function of DNA was at first unclear to research scientists, but by the 1950s, this enigmatic substance had turned out to have the biggest and most complex molecules found in living

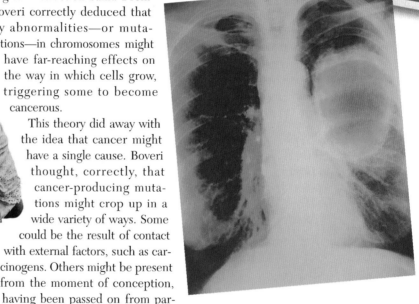

things. Furthermore, DNA was discovered in all living cells, and always found in rigidly fixed amounts—exactly what would be needed of something that stores the chemical instructions that make cells work.

The first hint that this might be so had actually surfaced in the late 1920s, when a British researcher named Frederick Griffith had made a startling discovery while working with bacteria. The bacterium in question, *Streptococcus pneumoniae*, exists in two forms: one can cause pneumonia, and the other is harmless. The only visible difference between them is that the virulent form has a smooth surface, whereas the harmless one is rough. These characteristics are normally passed on faithfully when the bacteria reproduce, but in 1928 Griffith found a set

HOW DOCTORS CHANGED THEIR MINDS

UNTIL THE 1960s, TOBACCO COMPANIES TRIED TO PERSUADE DOCTORS AND THE PUBLIC THAT SMOKING WAS HARMLESS, AND SOMETIMES EVEN GOOD FOR HEALTH

In today's health-conscious times, most people know that smoking increases the risk of cancer and other serious diseases. But at the beginning of the century, when cigarette smoking was relatively uncommon, perceptions were very different. Writing in the *New York Times* in 1906, one doctor recommended smoking as a healthy way of overcoming the stimulating effect of tea. In 1918, a group of surgeons urged that American soldiers in Europe be given a plentiful supply of cigarettes, in view of their supposedly beneficial effects.

When reports in the 1920s first began to connect smoking with ill health, tobacco manufacturers quickly acted to head off medical concern. For example, they admitted that some cigarettes could cause "throat irritation," using this to distract attention from more dangerous side effects.

One of the earliest companies off the mark was American Tobacco, which in 1927 sent almost every doctor in the country a special pack of 100 cigarettes. American Tobacco's chief selling point was that their cigarettes were made of "toasted" tobacco, and a questionnaire accompanied the free

FALLING REPUTATION A Lucky Strike advertisement from the 1930s promotes smoking as an alternative to "overindulgence." Below, a woman reads a 1970s poster about the dangers of smoking while pregnant.

SMOKING AREAS As concern about passive smoking increases, designated smoking areas (below) and the no smoking sign (below right) are increasingly common features of public places in many parts of the world.

cigarettes, asking doctors if the toasting process was "likely to free the cigarette from the irritation to the throat." More than 20,000 doctors thought it was. From the 1930s to the 1950s, cigarette advertisements regularly appeared in medical journals, as tobacco companies sought to dampen any disquiet about the implications of smoking.

In the early 1950s, more rigorous scientific methods began to expose strong links between smoking and disease. Two Americans, Ernst Wynder and Evarts Graham, showed in 1950 that male lung cancer sufferers were more likely to be smokers than men picked at random. In 1954, British epidemiologists Richard Doll and Bradford Hill found that doctors who smoked heavily were 24 times more likely to die of lung cancer than doctors who did not smoke. By the early 1970s, the medical profession had adopted a firm anti-smoking stance, bringing it onto an inevitable collision course with the tobacco industry. Today, the concept of smoking as harmless recreation is long gone.

of circumstances in which the normal rules were broken. If he killed some virulent bacteria and mixed them with live, harmless ones, the live bacteria produced virulent offspring. The dead bacteria obviously could not breed, but even so, they had managed to pass on their characteristics. Griffith was fascinated by this process of "transformation," but he died before anyone could explain it.

In 1944, three American biochemists—Oswald Avery, Colin MacLeod and Maclyn McCarty—identified the transforming agent as DNA. They showed that fragments of DNA could seep out of dead bacteria, to be taken up by bacteria that were alive. Once absorbed, the fragments were built into the recipient's chromosomes, where their instructions could be put into effect. This showed beyond doubt that the DNA held chemical information—information that could be transferred from one generation to another when cells reproduce.

Once DNA had been identified as the storehouse of genetic information, work began on unlocking its chemical structure.

THE DANGERS OF ASBESTOS

At the beginning of the 20th century, asbestos seemed to have an exceptional future. It could be used to insulate pipes and electrical wires, to build roofs, to filter fluids and above all to protect against fire. By the late 1930s, it had found its way into thousands of different products from buttons to brake linings, and asbestos-related industries employed hundreds of thousands of people. But while the boom in asbestos use was under way, some disturbing facts began to come to light. During the early 1900s, factory inspectors in Britain noticed increasing health problems among people involved in the asbestos industry. Workers exposed to asbestos dust often developed a condition in which the inner lining of their lungs became thickened and scarred, and in some cases this led to mesothelioma, a form of cancer affecting the membranes that surround the lungs.

In Britain, regulations governing asbestos processing were introduced in 1931, but elsewhere the enthusiasm for asbestos remained undimmed. Public health officials were aware that high levels of asbestos dust could be dangerous, but thought that low levels were quite harmless.

By the late 1960s, doctors began to report cases of asbestos-related cancer in people who had come into contact with only small amounts of asbestos fibers. Statistical projections at the time were so alarming that some countries—including the United States—banned asbestos altogether. This heralded the start of a hugely expensive program of asbestos removal that affected millions of schools, hospitals and other public buildings throughout the world.

REMOVING ASBESTOS   Many experts think that leaving asbestos in place may be safer than removing it.

The results of that work—published by James Watson and Francis Crick in 1953—showed that each DNA molecule consists of two chemical strands that spiral around each other like an endlessly twisted ladder. The strands are held together by four different kinds of chemical bridges, and the exact arrangement of these bridges, which varies from one living thing to another, forms instructions in a chemical code. In isolation, the instructions are meaningless, but once they are decoded by a cell, they control every aspect of the way it lives. Crucially, in the case of cancer, that includes the rate at which the cell divides.

Oncogenes: Chemical time bombs

Soon after Watson and Crick's discovery, the battle against cancer opened up on a completely new front. Oncologists—or cancer researchers—moved on from studies of individual cells to look at individual genes. By comparing chromosomes from normal cells with ones in tumors, researchers in the 1960s began to identify a number of specific genes that seemed to be involved in cancer. In 1978, Robert Weinberg of the Massachusetts Institute of Technology triggered cancer in mice by transferring single genes from one animal to another. In the same decade, researchers discovered that

SPIRAL OF LIFE   James Watson and Francis Crick pose by a model showing the newly discovered structure of DNA.

some viruses can carry out an almost identical process themselves, when they infect living cells. Once they are inside their hosts, these viruses insert cancer-causing genes into their host's chromosomes, leaving the chemical equivalent of a time bomb that may be activated immediately or many years later. These microorganisms, called retroviruses, include the one that causes AIDS, and the one that caused Peyton Rous's chickens to develop cancer.

Remarkably, investigation of these viral time bombs, or "oncogenes," soon revealed that they are very much like some genes that already exist in animal cells. Researchers also found that oncogenes are of no specific use to viruses themselves—instead they are simply genetic hitchhikers, having come from animal cells in the first place. But why should normal cells have genes that can cause cancer? This disturbing question lies at the forefront of cancer research today.

It now seems likely that for every oncogene, there is a harmless "proto-oncogene" that plays an essential part in normal cell

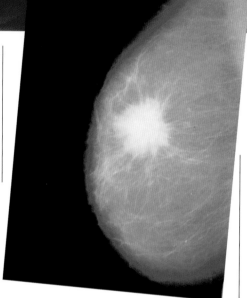

BREAST SCREENING Breast cancer still affects millions of women worldwide. Early detection is the best chance of successful treatment and scanners are widely used to screen suspected cases. A mammogram of a breast reveals the white mass of a cancerous tumor (right).

BREAKING THE SILENCE

In 1912, an American statistician named Frederick Hoffman calculated that cancer deaths cost his employer, the Prudential Insurance Company, more than $700,000 a year. Faced with a wall of silence that surrounded the disease, Hoffman decided to do something that would bring the problem of cancer into the open. In 1913, he helped to set up the American Society for the Control of Cancer to run public education campaigns about the disease.

growth and development. Unfortunately, these proto-oncogenes have a tendency to mutate into oncogenes, for example when they are exposed to a carcinogen, and it is then that the trouble starts. Once a mutation has occurred, the oncogene begins to issue inappropriate instructions, and the cell embarks on a pathway that can lead toward uncontrolled division. Such a process is remarkably close to Boveri's hypothesis, put

forward long before the science of molecular genetics was born.

At the beginning of the century, treatments for cancer were fairly few. One of the simplest—removal of the cancerous growth—already had a long history, and sometimes succeeded if it was carried out early enough. But success depended on removing all the affected tissue, together with any other tissue that cancerous cells might have reached. As the century progressed and more was discovered about the spread of cancer, ideas about how this should be done often changed.

Surgical treatments for cancer

One example of this process occurred with the treatment of breast cancer. In the late 1890s, William Halstead—the leading American surgeon who started the use of rubber gloves—devised a new operation for breast cancer called radical mastectomy. The operation was "radical" because it involved the removal not only of the breast itself, but also the connected lymph glands and muscles, which Halstead thought might harbor cancerous cells.

Such was Halstead's prestige that the new operation became a standard procedure for the next 70 years. Despite the drawback of disfigurement, the logic behind it seemed persuasive. In the late 1930s, however, surveys revealed that women who had less drastic surgery seemed to survive equally well—a finding that threw serious doubt on the routine use of the operation. Habits were slow to change, but by the 1960s radical mastectomy lost its place as the recommended form of treatment. It was widely supplanted by less intrusive surgery.

In the 1890s, when Halstead first developed his new operation, surgery of any kind was a risky business. Many kinds of cancer

THE EARLY DAYS OF RADIOTHERAPY

Radiotherapy works by bombarding malignant cells with high-energy radiation so that the cells cease to grow, or die altogether. In the early decades of the century, the most commonly used source of radiation was radium, a naturally radioactive element. But radium was exorbitantly expensive: in 1912, a gram could cost $180,000. At first, radium was often used by painting it onto stamp-sized metal plates, which were then strapped to the patient's body. Radiologists soon discovered a more useful form of radiation, which was created when radium underwent radioactive decay. Unlike radium, this substance was a gas. Called "radium emanation," it is now known as radon.

In radiotherapy centers, "radium emanation" was collected and sealed in hollow glass tubes and "seeds," which could be inserted into the body. Radium has a very long half-life, which meant that it had to be removed from the body once it had been used. Radon, on the other hand, has a half-life of just under four days, so once the glass tubes or seeds were implanted, their radioactivity soon fell away. As a result, they could be inserted into an affected area and then safely left in place.

In the early days of radiotherapy, the correct assessment of a radiation dose was a problem, and many patients suffered severe burns. One test for radiation strength used butter as a substitute for living tissue. A radon "seed" would be placed inside a butter pat, and a bleached zone in the butter would show how far the radiation had spread.

accepted by most cancer sufferers, but by the 1960s a change in attitude was emerging. Instead of automatically submitting to the surgeon's authority, patients demanded more of a say in the way they were treated. Today, doctors and patients alike are much more aware of the relative costs and benefits of surgery, and also of radiotherapy and chemotherapy, the two other forms of treatment in the battle against the disease.

X-rays against cancer

In 1903, just seven years after they had been discovered, a German surgeon named George Perthes found that X-rays could inhibit the growth of some kinds of tumors. Radiologists already knew—sometimes through personal experience—that X-rays could trigger cancer itself, but Perthes's breakthrough was the first indication that they could also have the opposite effect. It soon became evident that X-rays were not the only form of radiation that had anti-cancer properties. The radioactive

ZEROING IN X-ray treatment for cancer in 1917 (right), and treatment with a betatron in the 1980s (below), which focuses high-energy X-rays precisely.

element radium, which was discovered by Marie and Pierre Curie in 1898, also turned out to have the same paradoxical ability: in small doses it could provoke cancer, but in larger doses it could destroy it.

This discovery opened up a new form of treatment. In 1906, the world's first "radium institute" was planned in Paris, and by the beginning of the First World War, similar institutions had opened in many European cities. At a time when life expectancy and cancer mortality were both rising rapidly, X-rays and radium promised a new line of attack against an unyielding adversary.

At first, several practical difficulties had to be overcome. Early X-ray machines were easy to make but difficult to control. They often bombarded the body with too much

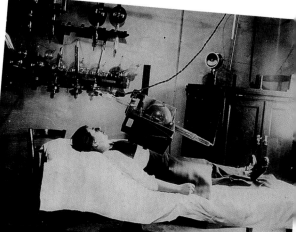

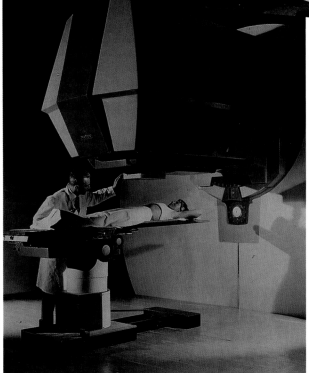

were inoperable, because removing them involved such a high risk of infection or blood loss that the patient was unlikely to survive. But with the introduction of blood transfusions in the late 1930s, and the subsequent discovery of antibiotics in the 1940s, surgery became much safer. There were now few places in the body where a surgeon could not attempt to seek out and remove a localized cancer.

These technical advances turned out to be something of a mixed blessing. Spurred on by their growing expertise, surgeons in the 1940s and 1950s became ever more ambitious in their battle with cancer. In some cases, new techniques dramatically increased the chances of long-term survival, but in others, drastic surgery did little to improve the patient's life. Often it was the surgeon, rather than the patient, who decided how far the disease should be pursued.

Initially, this kind of intervention was

radiation, or emitted it too feebly, failing to reach the target tissue. Radium, by contrast, was difficult to obtain and hazardous to handle. It took Marie and Pierre Curie four years to produce their original 0.003 ounce sample of radium, using several tons of ore. Even when the purification process became mechanized, radium was still very expensive and difficult to work with because, unlike an X-ray machine, it gives off radiation all the time.

Despite these problems, radiation treatment—or radiotherapy—soon had an enthusiastic following. By 1920, radiation treatment

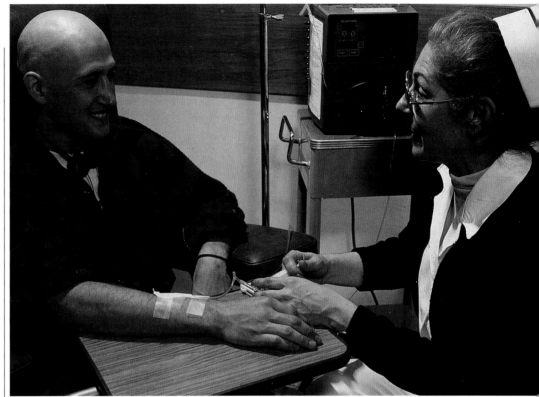

centers were a common feature of large hospitals in North America and Europe, and medical facilities became the largest market for radioactive elements. At a time before the development of atomic bombs and nuclear power, radioactivity was seen as something positive that had been discovered by modern science.

Some tumors turned out to be particularly responsive to radiation treatment, and radiologists experimented with new ways to deliver a lethal dose to malignant cells without harming too many of the healthy cells around them. One method, still in use today but with vastly improved technology, involved focusing an X-ray beam onto the tumor from a variety of angles. Another, widely employed during the 1920s and 1930s, used radium-bearing needles to deliver the radiation to exactly where it was needed. Again, this technique is still in use today, although radium has been replaced by safer elements such as cobalt or iridium, which are turned radioactive by artificial means.

By the middle of the century, the

PATIENT OBSERVATION

In 1978, 34-year-old Dorothea Lynch discovered that she had breast cancer. Her diary, illustrated with photographs by her friend Eugene Richards, was published as *Exploding into Life*. It is an account not only of her own treatment for cancer but also of the experiences of other patients in American hospitals.

"On the green TV screen above my head, I see my femur, my pelvis, even my liver, outlined by radioactivity, all looking like sparkling constellations. I have been injected with 'tracer material' and am lying on a hard table that moves up and down and sideways and makes little clicking sounds each time a picture is taken. This bone scan is a search for additional tumors, cancer seedlings, that may already have slipped like invaders from my breast to my lymph glands and into my blood.

"After an hour a small Japanese doctor appears next to me, beaming and clapping his hands together. 'Clear. Clear,' he says. 'One hundred percent.'

"My hands are shaking, and I try to catch my lower lip between my teeth to stop it from quivering. I have no idea what he means.

"'Good news,' the nurse says, squeezing my hands.

"Good news. There are no distinct metastases. The cancer is still in my breast, but it has not yet spread anywhere else. A little cancer, not a lot. I think they are telling me I am not dying."

strengths and weaknesses of radiotherapy had become clear. On its own, radiation often failed to clear up cancer, but when used in conjunction with other treatments, it reduced the cancerous cells' chances of survival. A three-pronged strategy, involving surgery, radiation and also chemotherapy, was now beginning to emerge.

Mustard gas therapy

By a curious twist of fate, the first real breakthrough in cancer chemotherapy came from a chemical with a sinister reputation. Dichlorodiethyl sulphide, better known as mustard gas, has horrific effects, as soldiers discovered in 1917. As clouds of the oily yellow substance blew over the trenches in northern Europe, frontline troops found that they could neither see nor breathe.

After the First World War, the full effects of mustard gas became apparent. It was found not only to injure the skin and lungs, but also to affect cells in bone marrow. A related compound, called nitrogen mustard, did the same thing, but was particularly toxic to cells that are replaced at a rapid rate. Prompted by this characteristic, a group of chemists at Yale University tested nitrogen mustard on mice in the 1930s. They found that it made tumors shrink rapidly; although the mice still died, they lived longer than they would have otherwise.

The problem of toxicity made doctors

SIDE EFFECTS A cancer patient receives an anti-emetic drug to help relieve the nausea caused by the chemotherapy that he is undergoing.

understandably reluctant to try nitrogen mustard on human patients, but during the Second World War, G.E. Linskog, an American surgeon, decided to risk its use. As the drug had potential military applications, it was known simply as "compound X," and Linskog had to guess the appropriate dose. His patient, who was suffering from poten-

MEASURING THE DOSE

Until 1928, radiotherapy involved a large amount of trial and error, because there was no internationally accepted unit of radiation dosage. That changed with the introduction of the roentgen, a unit of radiation named after Wilhelm Roentgen, who discovered X-rays in 1896. A further unit, the "roentgen equivalent man" or rem, took into account the effect of the radiation on the body. Today, these units have been largely superseded by the gray and the sievert.

tially fatal cancer of the jaw, initially responded well as the malignant growth shrank. But the respite was short-lived. The cancer eventually recovered its grip, and each successive dose of the drug proved less effective than the one before.

Despite its drawbacks, the use of nitrogen

mustard proved an important point. It showed that some drugs could affect cancerous cells while leaving healthy cells relatively unharmed. But even so, chemotherapy (treating the disease with chemicals) was a blunt-edged weapon. Any drug powerful enough to kill cancerous cells outright was almost certain to kill other cells as well.

Since the experiments with nitrogen mustard, nearly half a million drugs have been tested for anti-cancer effects. There have been some valuable discoveries—including alkylating agents related to mustard gas, antibiotics that can kill proliferating cells, and also substances called vinca alkaloids, found in periwinkle plants. However, the number of useful drugs remains disappointingly small.

Over the years, the search has collected its share of medical controversy, with hotly disputed claims being made for some substances that have failed official tests. One of these, called laetrile, caused a lengthy skirmish in the 1960s and 1970s, when it was championed by enthusiasts in the United

States, but branded as worthless by the Food and Drug Administration. Laetrile is made from apricot kernels, and was eagerly sought after by many cancer victims, despite the lack of any scientific evidence that it worked. The Food and Drug Administration eventually banned it from interstate sale. Two decades later, laetrile is rarely used.

There have been some successes in different kinds of chemotherapy, based on substances naturally produced by the body. Sex hormones, and substances that block their production, have proved useful in controlling the development of breast cancer and cancer of the prostate gland. Substances called interferons, which are produced by healthy cells when attacked by viruses, have been used in the fight against leukemia and some kinds of skin cancer. In conjunction with surgery and radiotherapy, chemotherapy provides a third line of attack, but after decades of research it still has drawbacks and is rarely used on its own.

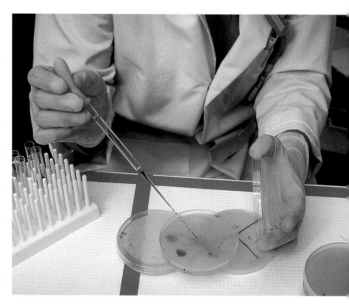

CANCER IN THE LAB Cancerous cells cultured in a laboratory dish are tested with an experimental drug.

The cancer research industry

At the beginning of the century, cancer research occupied a modest position in the medical world. Since then, it has grown into a huge enterprise. It has generated careers, alliances and rivalries, and it has given politicians an unprecedented opportunity to steer medical research—often against the judgment of researchers themselves.

One of the most celebrated examples of this kind of intervention occurred in the United States in 1971, at a time when President Nixon was seeking re-election. Nixon had been instrumental in creating the National Cancer Act, a piece of legislation that created a "war on cancer," the implication being that with some determination and persistence, victory lay not far off. The Act established a five-year battle plan against the disease, with a series of high-level conferences involving leading

NUCLEAR FALL-OUT Ten-year-old Oxana has a cancerous tumor in her mouth, caused by radiation from the 1986 Chernobyl explosion.

researchers in the field. The president believed that such a bold initiative could only bring credit to himself.

Even though many researchers were skeptical about joining a scramble for victory, the "war" was lavishly funded. The budget of the National Cancer Institute rose from $1.75 million in 1946 to $400 million in 1973, and then to more than $1 billion during the following decade. But while this huge investment captured the headlines, it produced relatively few returns. By the 1980s, when Nixon was no longer on the scene, the battle plan had come undone. Critics of cancer research, in America and elsewhere, voiced the growing conviction that money alone was unlikely to provide some all-encompassing cancer "cure."

As the century closes, the business of cancer research is still one of the most important and generously funded branches of medicine. However, there is now agreement that the answer to cancer lies not solely in the laboratory. It also lies in environmental and lifestyle changes aimed at cancer prevention, and in improved diagnostic services—such as breast X-ray screening—that help to detect cancer while it is still in an early stage.

In the future, cancer research may produce some truly revolutionary discovery, enabling cancerous cells to be inactivated before they do any damage. But until that happens, the experiences of the last 100 years show that prevention will remain at least as important as the search for a cure.

THE HEALTHY MIND

CENTURY OF PSYCHIATRY HAS PRODUCED NEW WAYS OF TACKLING MENTAL ILLNESS, BUT THE CAUSES OFTEN REMAIN ENIGMATIC

In 1960 Thomas Szasz, a Hungarian-born psychoanalyst trained in Chicago, published a book called *The Myth of Mental Illness*. In it, he argued that mental illnesses had nothing in common with organic diseases, such as heart failure or the common cold. Instead, they were medical inventions—devised by doctors and psychiatrists to label conditions with no physical basis.

At the time, Szasz's book had a profound impact on doctors and psychiatrists—particularly in America—because the causes of mental illness were a continuing puzzle. After five decades of extraordinary advances in other fields of medicine, mental disorders continued to perplex. It was true that a handful had been linked to physical causes: as early as 1906, for example, Alzheimer's disease was linked to brain-cell changes, and paresis, or general paralysis of the insane, was shown a few years later to be an effect of syphilis. But by the 1960s, most forms of mental illness, including schizophrenia and manic depression, still had no apparent biological cause. To many readers of Szasz's book, the idea that these disorders were

purely psychological conditions brought about by the pressures of human existence was highly persuasive.

Today, Szasz's ideas have lost much of their original scope. Since the 1960s, evidence has accumulated that some mental illnesses have a chemical basis, which in some cases can be inherited. But despite these discoveries, many mental disorders remain difficult to explain.

The rise of psychoanalysis

At the beginning of the 20th century the science of psychiatry had reached a crossroads. In Germany and Austria, psychiatrists had made great progress with mapping the functional areas of the brain, but the hunt for abnormalities that could be linked to disease had yielded few results.

Despite this inconclusive search, most researchers saw mental disease in physical terms. Many still believed in "degeneration," a late 19th-century theory developed by Bénédict-Auguste Morel. Morel worked in a number of French asylums, and became convinced that mental disorders were often

passed on from one generation to the next, becoming more pronounced in the process. No clear evidence for degeneration was actually found, and as the new century got under way, disenchantment with the biological approach to mental disease set in.

It was at this point that the most remarkable and controversial figure in 20th-century mental health took the stage. From his office in Vienna, the neurologist Sigmund Freud undertook consultations with patients suffering from neuroses, or behavioral disorders, that can make everyday life difficult without actually making it impossible. During the course of this work, Freud encouraged his

ALL IN THE MIND

For most of the 20th century, stomach ulcers were thought to be a product of high-stress lifestyles. But in the 1980s, researchers found that most people suffering from ulcers harbor an infectious bacterium called *Helicobacter pylori* in their stomachs. It now seems likely that these bacteria cause ulcers, although stress may play a part in allowing them to grow.

patients to reveal their innermost thoughts as candidly as possible. Initially he used hypnotism, but he later replaced this with a technique called free association, which helps bypass the checks that are normally maintained by the conscious mind. Freud's conclusions were remarkable and, to many people at the time, also shocking. He found that neurotic behavior was the result of repressed desires and experiences, primarily sexual in origin, that dated back to early childhood.

This technique, which Freud called psychoanalysis, gave birth to a form of treatment called psychotherapy. In its original form, as practiced by Freud, the patient would visit the therapist five days a week for months or even years, and a relationship of complete trust would develop. Guided by the therapist's detached but probing questions, the patient would gradually reveal his or her most deep-seated fears and preoccupations. This would allow the patient to examine feelings

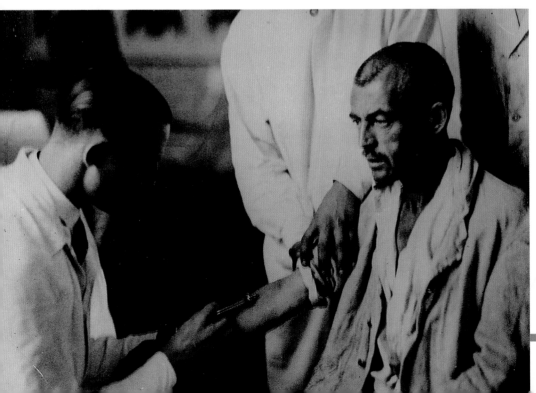

SHOT IN THE ARM A German psychiatric patient receives morphine in the early 1930s. Despite being addictive, the drug was used to subdue agitated patients.

WATER TREATMENT Wearing inflatable pillows, two Russian patients relax during a hydrotherapy session at a Moscow hospital.

that previously had remained in the subconscious, helping to erase the repression that was the cause of the original neurosis.

Freud himself made relatively few claims about the therapeutic value of his technique, but his followers were less restrained. By the 1920s, psychoanalysis had become highly fashionable among middle-class Austrians and Germans, and psychotherapy was increasingly accepted as a way of dealing with behavioral disorders. There were many skeptics—one German professor of psychiatry, for example, denounced the psychoanalytic theory as being nothing more than a "triumph of suggestion"—but the tide of opinion was against him. In an increasingly polarized profession, psychoanalysts and "biological" psychiatrists clashed over the nature and treatment of mental disease.

Freud visited the United States in 1909, and the nation proved to be a fertile seedbed for his ideas. In the 1920s and 1930s, more and more American psychiatrists took up private practice as psychotherapists, and a wave of Jewish emigration from Germany and Austria during the early Nazi era added prominent practitioners to their numbers. By the 1950s, the United States had taken over from Germany as the leading center of psychiatry. Biological psychiatry was in the doldrums, and psychotherapy—commonly known as the "talking cure"—reigned supreme.

Despite its widely publicized successes, psychotherapy had its limits. It could help people suffering from neurosis, but it rarely helped people suffering from psychosis, the catch-all term used to describe more serious forms of mental illness. Faced with a lack of practical treatment, most psychotic patients ended up as long-term inmates of special mental hospitals, where any form of therapy was rare. As the decades went by, some of these hospitals grew to an enormous size. In the 1950s, for example, the largest mental hospital in the country, at Milledgeville in Georgia, housed more than 10,000 patients.

A series of shocks

In institutions like these, inmates were a world away from the comfortable consulting rooms of private practice, with their unhurried exchanges between therapist and patient. But mental hospitals were not always places where the sick were simply shut away and forgotten.

In a few hospitals, important new forms

GENIUS OR FRAUD? More than 50 years after his death, Freud remains the most controversial figure in 20th-century psychiatry.

1909 Sigmund Freud visits the United States

1917 Wagner-Jauregg induces malaria fever in a patient

1926 Otto Loewi discovers the role of neurotransmitters

1930s Birth of modern psychosurgery

1939 First use of ECT on a human patient

1943 Manfred Sakel claims success for insulin-coma therapy

of treatment were pioneered. One of the earliest 20th-century discoveries came with the work of Professor Julius Wagner-Jauregg, an Austrian psychiatrist who worked in Vienna. In the 1880s, Wagner-Jauregg noticed that a psychotic female patient had regained her senses after catching an infec-

BRAIN WAVES

The use of electrodes to monitor the brain's electrical activity was first described by the German psychiatrist Hans Berger in 1929. The instrument that he pioneered was called the electroencephalograph or EEG, and revealed the existence of "brain waves," or rhythmically changing patterns of activity. Abnormal EEG readings proved a useful way of diagnosing epilepsy, and of locating brain damage.

tion. He suspected that there might be a connection between the two events, and set about trying to establish if this was true.

During the early 1900s, Wagner-Jauregg managed to help patients suffering from syphilitic paresis, or general paralysis of the insane, by injecting them with a tuberculosis

PRIVATE LIVES The spartan interior of a psychiatric clinic in Naples, 1979. In many countries, mental illness is still seen as a matter for secrecy and shame.

vaccine. The procedure was partly successful, and Wagner-Jauregg concluded that the fever produced by the vaccine had triggered the improvement. In 1917 he deliberately injected a syphilitic 37-year-old actor with the parasites that cause malaria, knowing that the fever produced by malaria would be much more intense than the one produced by the TB vaccine. Six months later, the former terminally ill patient was discharged as fit and well.

Wagner-Jauregg's fever treatment triggered a wave of interest in physical therapies for people with severe mental illness. He had used fever to shock the brain out of its abnormal state, and a search now began for other physical shocks that might have similar useful effects. In the early 1930s, Manfred Sakel, a Jewish psychiatrist working in Vienna, tried a very different procedure. Using the hormone insulin, he put his patients into a coma.

Insulin-coma therapy was a risky undertaking. Once injected, the insulin lowered the patients' blood sugar level until there was not enough left to maintain consciousness. The patients would then be left hovering on the threshold of respiratory failure for up to 20 minutes, until a sugar solution poured through a nasal tube brought them back to life. This process was repeated a dozen or

FEVER TREATMENT Experiments conducted by Julius Wagner-Jauregg showed that some forms of mental illness could be combatted by deliberately provoking fever.

more times, each coma being deeper than the one before. The amount of insulin needed was difficult to gauge, and in about one case in a hundred, the patient would fail to regain consciousness and die.

The new form of therapy produced remarkable results. In 1943, Sakel reported that 70 percent of his schizophrenic patients experienced full remission of their symptoms, and nearly 20 percent experienced partial remission.

Despite its considerable dangers, insulin therapy did find a place in many mental hospitals. So too did an even more alarming form of therapy that used a drug called metrazol. When administered in large doses, this substance, which had originally been formulated as a heart stimulant, produced convulsions rather than coma. Over a course of several treatments, these convulsions seemed to turn seriously disturbed patients into normally behaved people. However, metrazol had a terrifying side effect: just before producing convulsions, it triggered a sense of overwhelming and almost primeval fear, known as a "metrazol storm." Many psychiatrists found this effect uncomfortable to witness; just how patients felt is difficult to imagine.

Ultimately, fevers and coma therapy did not stand the test of time. However, there is one form of shock therapy dating back to the 1930s that is still used today—electroconvulsive

1951 Henri
Laborit's work
on 4560RP

1960 Thomas Szasz
publishes *The Myth
of Mental Illness*

1963 Valium
first
marketed

1970s Fluoxetine, now
known as Prozac, first
synthesized

therapy, or ECT. During ECT an electric shock is passed through the patient's head between two electrodes placed on the temples, causing loss of consciousness and convulsions. It is still used in psychiatric medicine, although after 60 years no one has any clear idea how it works.

Brain surgery

In the 1930s, two American researchers conducted an operation on a chimpanzee, destroying the frontal lobes in its brain. The chimp, which had previously been noisy and aggressive, became quiet and docile. On learning about the result of this operation, a Portuguese neurologist named Egas Moniz pursued the discovery to its logical conclusion, and in 1935 persuaded a surgeon to carry out a similar operation on human patients. With this development, modern psychosurgery was born.

Moniz called his operation a leucotomy, although it eventually became better known by the American term lobotomy. He claimed that it produced immediate improvements in a large proportion of patients and that, despite its drastic nature, it had few harmful

DEFINING MENTAL ILLNESS

In 1968, the American Psychiatric Association published the second edition of its *Diagnostic and Statistical Manual of Mental Disorders*. It listed all the mental abnormalities that were recognized at the time—180 in all. When the third edition was published in 1980, the number of recognized disorders had grown to 265.

side effects. News of Moniz's breakthrough was welcomed by the American neurologist Walter Freeman, and from the mid-1930s he and his colleague James Watts became keen practitioners of the operation.

In their original version of the lobotomy,

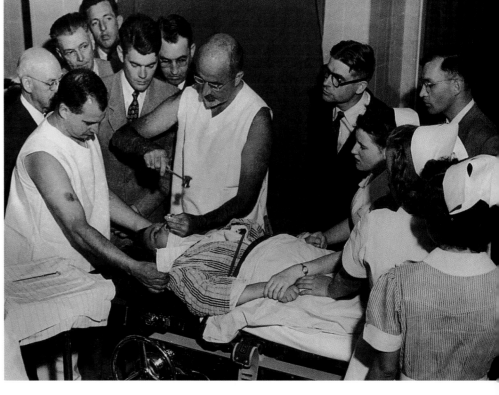

Freeman and Watts used a T-shaped instrument called a "precision leucotome." This had a long blade that was inserted through a small hole in either temple. The blade would then be levered up and down, severing the connections between the frontal lobes and the rest of the brain. In *Psychosurgery*, which was published in the late 1930s, the result was illustrated with X-ray photographs. These showed that despite the accent on "precision," the leucotome actually cut a jagged path through the brain, destroying a large amount of tissue in the process.

Remarkably, this kind of wholesale damage seemed to produce few wholly negative results. Restless or violent patients were often more subdued after surgery, encouraging Freeman in particular to promote the operation. Watts became doubtful about the wisdom of the technique and parted company with his colleague

UPPER CUT Watched by a small but attentive crowd, Walter Freeman hammers a lobotomy knife through a patient's eye socket. The patient is fully conscious.

in 1947, but from that point on, Freeman energetically set about turning lobotomy into what he called an "office procedure." He developed a new form of the operation named transorbital lobotomy, which involved hammering a blade through the skull just behind the upper eyelid. The operation was quick and easy, and Freeman carried it out across the country, packing his equipment in the back of a station wagon.

On both sides of the Atlantic, lobotomy enjoyed a period of acceptance as a new way of dealing with difficult patients. But, as was also the case with less extreme forms of physical therapy, a string of negative side effects eventually became apparent. Lobotomized patients were certainly more tranquil, but as well as losing their anger, fear or agitation, they often lost much of their personality as well. By the 1950s, lobotomy looked more like a way of making life easier for caregivers than a genuinely therapeutic measure.

Despite its fall from grace, Freeman's operation continued to be used on

LOBOTOMY INSTRUMENTS The long blades on these knives were designed to slice sideways through several inches of brain.

CHANGING THE BRAIN

In their book *Psychosurgery*, published in the 1940s, Walter Freeman and James Watts included transcripts of conversations between surgeon and patient during lobotomy operations. Because the brain has no nerve endings and cannot feel pain, most patients were given an external local anesthetic, enabling the surgeon to assess the result of his handiwork while the operation was in progress.

In these chilling extracts from case number 285, "Frank," a 24-year-old schizophrenic, shows the effects of the lobotomy as his emotions are literally cut away:

"9:30 a.m. Towel sutures going in. Frank is quiet except for grunting and groaning. He was alert and smiling earlier, but now he is tense and overventilating; his hands are cold and clammy. Pulse 68; blood pressure 102/56. He repeats over and over: 'Oh, I'm going, I'm going. I can't breathe.'

Doctor: Are you scared?
Patient: Yeah.
Doctor: What of?
Patient: I don't know, doctor.
Doctor: What do you want?
Patient: Not a lot. I just want friends. That's all. How long's this going on?
Doctor: Two hours.
Patient: Two hours? I can't last that long.
Doctor: How do you feel?
Patient: I don't feel anything but they're cutting me now.

The account records that a number of sweeping cuts are now made through the patient's brain. "Frank" becomes alarmed and asks the doctor to "stop experimenting," but the operation continues. Further cuts are made, followed by a left and right stab. The patient jerks his head, but after the right-hand stab, his voice becomes muffled:

Doctor: Are you uncomfortable?
Patient: No.
Doctor: Why do you jerk around?
Patient: I don't know.
Doctor: Can you breathe?
Patient: Yes.
Doctor: Feel fine?
Patient: Yes, pretty good.

The lobotomy complete, the surgeon checks his patient's mental state.

Doctor: What place is this?
Patient: George Washington Hospital.
Doctor: What are you doing here?
Patient: Being operated on.
Doctor: What for?
Patient: There's something wrong with the brain. I don't know what it is.
Doctor: Is it all right now?...
Patient: Yes. I feel pretty good right now."

---

a greatly reduced scale until the 1960s. By then, pharmacology had entered the world of psychiatric medicine, and a series of developments in the use of drugs relegated lobotomy to the sidelines. The entire surgical excursion became an episode most psychiatrists were more than happy to forget.

Drugs and the mind

In 1926, while psychoanalysis was starting to make inroads into professional psychiatry, an Austrian pharmacologist named Otto Loewi made a momentous discovery. Nerves were known to work by electrical impulses, but Loewi found that a chemical called acetylcholine was involved in transmitting impulses from one nerve cell to another. Acetylcholine, and substances like it, became known as neurotransmitters.

Some psychiatrists tried using acetylcholine as a therapeutic drug—without any success—but Loewi's discovery opened the way toward a wider search for "psychotropic," or brain-influencing chemicals. The first real success did not come until shortly after the Second World War, and it happened largely by accident. John Cade, a medical officer in a mental hospital in Australia, was investigating substances in the urine of manic patients, in the hope of finding some chemical abnormality. He injected the urine into guinea pigs, and used the element lithium to make some of the substances in urine more soluble. To his surprise, Cade found that lithium itself had an extraordinary effect on the rodents. Instead of struggling and squeaking noisily as guinea pigs normally do when handled, they became so placid that they could be laid on their backs.

Excited by this turn of events, Cade courageously tested lithium on himself, and then gave it to a number of his patients. It had no effect on patients who were depressed, but its impact on manic patients exceeded his most optimistic hopes. All of them improved, and some were soon fit enough to be released.

Cade's discovery proved to be the earliest in a rapidly expanding field. In the late 1940s, as news of lithium treatment spread, Henri Laborit, a French naval surgeon, was experimenting with drugs that could reduce post-surgical shock. When he tested a group of chemicals called phenothiazines, he noticed that patients undergoing operations became strangely calm and relaxed. For a

---

while, Laborit did not pursue the matter, but in 1951, when transferred to the Val-de-Grâce military hospital near Paris, he carried out research on one particular phenothiazine, called 4560RP. It was effective in the operating room, but this was eclipsed by its psychiatric value. 4560RP could calm the most manic and violent of patients, without

CHEMICAL LINK Otto Loewi's discovery of acetylcholine changed ideas about the way the nervous system works.

any of the dangers associated with shock therapies. It was soon being used in mental hospitals throughout France.

The drug 4560RP became better known as chlorpromazine or Largactil. As its use spread to other countries, extraordinary scenes took place. Long-term patients were able to return to normal life, and many were reunited with former partners who had given them up as lost.

It was not appreciated at the time that chlorpromazine did have its dangers—a series of side effects, such as lethargy and mental vacancy, that built up with extended use. But the advent of this new drug marked an important turning point in 20th-century

ELECTROCONVULSIVE THERAPY

SIX DECADES AFTER IT WAS FIRST TRIED OUT, ELECTRIC SHOCK TREATMENT CONTINUES TO PLAY A CONTROVERSIAL ROLE IN THE TREATMENT OF MENTAL ILLNESS

On April 15, 1938, a 39-year-old schizophrenic found wandering about in a train station in Rome was detained by the police, and sent to a local mental hospital "for observation." Three days later, the man became the first patient to undergo a form of treatment called electroconvulsive therapy, or ECT. Electrodes were placed on his temples, and a rubber chock inserted between his teeth. One of the staff threw a switch, and a burst of electricity surged through the patient's head.

The first dose seemed to have little effect, apart from making the man's muscles jolt. A second was given, this time at a higher voltage, and the patient began to protest. But after the third dose, when the voltage was turned up to the maximum, the man underwent a convulsion. His breathing stopped, his reflexes came to a halt, and he stayed in this state for nearly a minute. Finally, he resumed breathing, and when he regained consciousness he seemed remarkably calm and collected. The treatment appeared to have shocked him into his senses.

ECT was developed by Ugo Cerletti, a professor of psychiatry, and his assistant Lucio Bini. Their research was prompted by the work of Ladislas von Meduna, a Hungarian psychiatrist, who had discovered that drug-induced convulsions could cure patients suffering from severe mental illness. However, the drugs used produced unpleasant or dangerous side effects, and the Italian team decided to find out if electricity offered a safer alternative.

Their first patient received nearly a dozen ECT treatments, and although not completely cured, he left the hospital in remarkably good health. News of "electroshock" therapy spread, and ECT departments opened in mental institutions worldwide. The treatment was helpful in schizophrenia, and particularly for dealing with depression although it could cause memory loss. In a refinement of the technique, patients were given a muscle relaxant before treatment to prevent sudden movements that could cause broken bones.

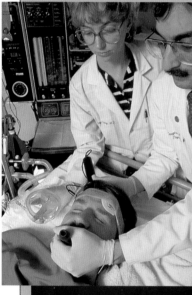

WIRED UP Shock therapy in an American hospital (left) compared with more primitive conditions in Romania.

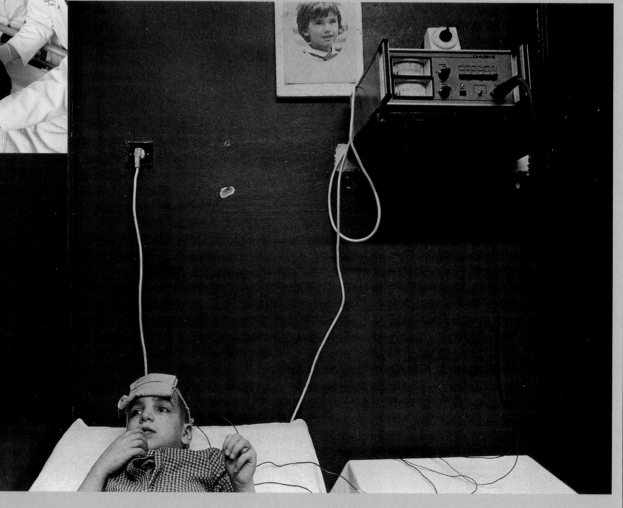

care of the mentally ill. In the mid-1950s, the number of patients in institutions at last started to fall as medication began to take the place of confinement.

From chlorpromazine to Prozac

Following the discovery of chlorpromazine, drug companies poured their resources into finding new psychiatric drugs. Initially, the search was for drugs that would help the severely mentally ill. But in the mid-1950s, with the discovery of tranquilizers such as meprobromate, or Miltown as it was known in America, medically approved mood-altering drugs entered daily life.

According to one estimate, by 1956 one in 20 Americans was taking tranquilizers on a regular basis. During the early 1960s, meprobromate was joined by a group of drugs called benzodiazepines, which relieved anxiety without causing heavy sedation. Anxiety—a state that was previously thought of as a normal part of life—gradually came to be seen as a medically treatable condition.

ONE OF MANY A patient rests in an Ohio mental asylum in 1946. Because of advances in medication, psychiatric patients are no longer confined in large numbers.

The most important of these new drugs, Valium, was first marketed in 1963. It shot to the top of the list of pre-scribed drugs in the Western world, and was soon being taken on a continuous basis by hundreds of thousands of people. It took more than a decade for doctors to realize that Valium was "psychologically addictive." This flood of drugs had some unsettling effects on the psychiatric profession, which by now relied on psychotherapy as a main form of treatment. Tranquilizers and "anxiolytics" such as Valium produced rapid results, while psychotherapy could take years to achieve the same effects.

In America, matters came to a head during a court case in the early 1980s, when a hospital that used psychoanalytic techniques was sued by one of its patients, a doctor named Rafael Osheroff. Dr. Osheroff had been suffering from depression, and was admitted for a stay that ultimately lasted seven months. At the end of this period, a frustrated Dr. Osheroff claimed that he had

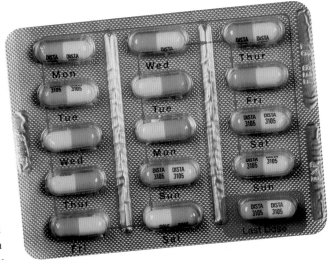

THE PROZAC AGE First made in the early 1970s, fluoxetine—or Prozac—is the latest in a line of psychotropic or mood-altering drugs.

not received appropriate treatment, and transferred himself to another clinic where he was given antidepressant drugs. He was soon able to return home, only to find that his wife had left him, and his professional standing had collapsed. Dr. Osheroff won his case, giving chemical therapy a boost.

The final two decades of the 20th century yielded more evidence that chemistry links the mind and the brain. Since the discovery of acetylcholine, intense research into other neurotransmitters has revealed a network of chemical intermediates that seem to affect human behavior and mood. One of these neurotransmitters—serotonin or 5HT—orig-inally discovered in the 1940s, is now thought to be implicated in sensory percep-tion and mood changes, and many drugs have been devised with the aim of increasing its levels in the brain. The most successful of these drugs, called fluoxetine, was first syn-thesized in the early 1970s. Now known by the trade name Prozac, the medication has acquired a Valium-like reputation as a won-der drug that can make depressed people feel "better than well."

In the world of severe mental illness, progress has also been made. The notion that psychosis is simply a reaction to life's pressures has buckled under evidence that some disorders, such as schizophrenia, do have a genetic link. Despite these findings, the mechanisms that trigger most psychiatric problems are still unknown. At the dawn of the 21st century, few psychiatrists think that biology alone will supply the answers.

ENJOYING LONGER LIFE

INCREASING LONGEVITY HAS BROUGHT WITH IT MEDICAL PROBLEMS AND DILEMMAS THAT WERE UNKNOWN JUST A FEW GENERATIONS AGO

The Russian revolutionary Leon Trotsky wrote that: "Old age is the most unexpected of all things that happen to a man." In Trotsky's case, it turned out to be a brief experience, because at the age of 72 he fell victim to an assassin in Mexico City. However, his words capture a paradoxical feature of the aging process: although it is something that most people experience, its effects often come as a surprise.

During the 20th century, an increasing proportion of the world's population survived to experience these effects for themselves. The maximum human life span—about 90 years, although over 120 years has been recorded—does not seem to have changed, as a look at old gravestones shows. But since 1900, life expectancy—a measure of how far the average person gets along the path to that maximum span—has increased by about

50 percent in most of the developed world. As a result, old age has become a much more significant part of life. With this has come a growing medical interest in the process of aging, and a realization that in the not-too-distant future, the old and very old will make up a substantial segment of society.

Until the middle of the 19th century, old age—rather like childbirth—was something that many doctors passed over as a non-medical matter. By the 1890s, the average family doctor would certainly have helped his elderly patients lead as comfortable an existence as possible, but the medical problems associated with old age still received little attention. Instead, it was accepted that with aging came inevitable deterioration, something that no amount of professional skill could prevent. The specialized study of medical care for the elderly did not yet exist, and the term that describes it, geriatrics, did not gain widespread currency until the new century began.

Geriatric medicine initially developed in the United States, and was largely the creation of one man, an Austrian-born doctor named Ignatz Nascher. Nascher studied medicine in New York, and during his period as a student he heard one elderly woman telling a professor about her problems. When he later asked the professor what could be done for the patient, the answer was brief: nothing at all. Nascher was so dismayed by this dismissive attitude that he set out to change it.

In the early 1900s, the subject of aging was entirely absent from the curriculum of medical schools, a situation that was to remain largely unchanged until the Second World War. However, despite this apparent lack of progress, Nascher did manage to put geriatric medicine on a scientific footing. He published the first comprehensive textbook on the diseases of old age in 1914—although only after a struggle to find a publisher—and he began a geriatrics section in a leading journal of medical reviews (the title, appropriately, was flanked by a pair of hour-glasses). He also wrote articles for popular magazines about the health problems of

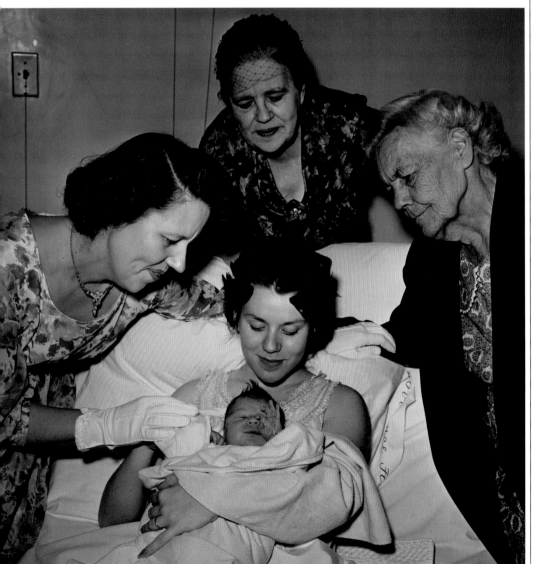

FIVE GENERATIONS Newborn Mercie McDuffy, held by her mother, is surrounded by her grandmother, great-grandmother and great-great-grandmother in a Washington, D.C., hospital in 1961.

1906 Alzheimer's disease identified

1914 Ignatz Nascher publishes the first textbook on the diseases of old age

1920s Serge Voronoff's "rejuvenating" operation for men becomes popular

1940s Research by C.M. McKay links aging to diet

1948 Hench and Kendall show that cortisone can treat rheumatoid arthritis

aging. This foray into the world of mass-market publishing was a particularly notable development, because Nascher was one of the few writers in the field. At the time, many doctors wrote articles for the general public—just as they do today—but hardly any tackled issues raised by aging.

The diseases of old age

At the height of Nascher's career, many conditions that are now recognized as specific diseases were thought to be simply the normal and inescapable effects of aging. Only one disorder linked with old age—stroke—had received any detailed medical

THE SOCIALIST CENTENARIANS

Who are the longest-lived people on Earth? At one time the answer—according to former Soviet politicians—was the native population of the Republic of Georgia, now an independent state. In the 1950s, Soviet gerontologists studying longevity focused on Georgia, where a group of people called the Abkhazians were reputed to live to immense ages. Research began to reveal villages where many inhabitants were over 100, and some were over 150. This longevity was often presented as a beneficial side effect of Communist life. However, since the collapse of the Soviet system, facts have emerged that cast doubt on the original claims. Most telling of all is that none of the "socialist centenarians" were able to produce birth certificates, as a system of birth registration only began during the Communist era. Evidence of age was based almost entirely on physical examination and the testimony of the centenarians themselves. Rumors suggest that some of the centenarian men may have assumed their fathers' identities in order to avoid being drafted into the Red Army, thus "aging" an entire generation overnight. There are few doubts, however, that the Abkhazians and other peoples in this part of southern Central Asia do live for an unusually long time, probably due to their diet and lifestyle: a plentiful intake of grains, fruit and vegetables, meat in moderation, and outdoor work in clean mountain air.

LONG LIFE Three generations of a Georgian family prepare an open-air lunch.

BACK INTO ACTION A stroke victim in Sun City, Arizona, pedals his way back toward a fully active life.

research as an illness in its own right.

Strokes are the result of an interruption in the brain's blood supply, usually as a result of a blood clot lodging in an artery. Like many of the diseases of aging, the incidence of strokes climbs steeply beyond the age of about 50. Strokes affect only about 0.1 percent of people between the ages of 55 and 65, but after the age of 75 the incidence is 20 times higher.

Strokes can often recur, and frequently increase in severity. However, until the beginning of the century, the only warning that a person was likely to suffer strokes was the first stroke itself. This situation changed with the advent of the first practical and accurate sphygmomanometer, a device that enabled doctors to make routine measurements of blood pressure. High blood pressure, or hypertension, increases the risk of strokes, so it can be used as a diagnostic test before trouble actually begins.

Until the 1920s, treatment for strokes was still fairly rudimentary, and consisted primarily of helping victims cope with their disabilities. Then, just before the 1920s began, a natural blood anticoagulant was discovered in animal livers and lungs, and was purified for use in human patients. Called heparin, it helped to dissolve existing clots and, if taken regularly, to stop them from forming. For the first time, doctors had a preventive medication that could do something to fend off one of the problems of old age.

Since the discovery of heparin, several other anticoagulants have been identified.

1961 First trials of L-dopa, a drug for sufferers of Parkinson's disease

1960s The anticoagulant Warfarin is given to potential stroke victims

One of the most important, a substance called Warfarin, was first synthesized in the 1940s. Despite its military-sounding name, warfarin has nothing to do with warfare—instead, its name is an abbreviation of the Wisconsin Alumni Research Foundation, the body that brought about its discovery. The Wisconsin team found that Warfarin acted more slowly than heparin, which made it more suitable as a long-term preventive medicine. Warfarin became widely prescribed for patients liable to strokes, and from the 1960s on, also for those who had heart surgery. Like all anticoagulants, it had to be used with care because it could cause uncontrolled bleeding. Patients who were prescribed Warfarin were often alarmed to discover that because of this effect the same substance—albeit in much stronger doses—was used as a rat poison.

With improved imaging techniques, such as the X-ray cross-sections of the brain provided by CAT (computerized axial tomography) scans, doctors were able to pinpoint the site of a stroke, and to identify places where future strokes might occur. During the last two decades, blood-vessel surgery has been used to minimize the risk of further attacks, but because brain cells are dependent on a constant supply of oxygen, there is still no method of bringing stroke-affected cells back to life. Today, as at the beginning of the century, recovery from a stroke is often only partial and requires patience and determination.

Arthritis

Arthritis is one of the most significant disorders affecting the elderly, and also one of the most ancient medical problems that we know about. Almost everyone over the age of 60 experiences some problems with their joints, but the symptoms vary. In some people arthritis means minor aches and

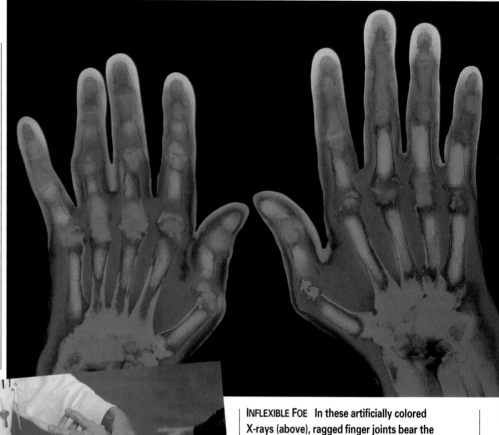

pains; in others, it leads to crippling deformities.

The most common form of the disease, osteoarthritis, seems to be a result of general wear and tear. In the late 19th century, doctors were familiar with the disease through their studies in the dissecting room, but with the discovery of X-rays in 1896, the effect on living joints became clear. X-rays also allowed doctors to examine a less common but more serious condition, rheumatoid arthritis, which occurs when the immune system attacks the tissues in and around joints. Like osteoarthritis, this disease usually develops in middle

INFLEXIBLE FOE In these artificially colored X-rays (above), ragged finger joints bear the imprint of rheumatoid arthritis, which commonly attacks hand and wrist joints (left).

age, but it can also erupt during childhood.

Although a lot of research has been devoted to preventing arthritis and minimizing its effects, successes have been relatively scant. One of the few specific breakthroughs came in 1948, when two Americans, Philip Hench and Edward Kendall, showed that cortisone, a hormone discovered in the 1930s, could be used to treat rheumatoid arthritis. Hench and Kendall received Nobel prizes for their work, but their discovery turned out to be less far-reaching than was hoped. Cortisone was helpful as a short-term measure, and is still used today, but subsequent research showed that it did not halt the progress of the disease. It sometimes produced high blood pressure and diabetes, and clinical tests showed that long-term treat-

INFORMAL NOTE Edward Kendall and Philip Hench pioneered the use of cortisone to treat arthritis.

ment with cortisone suppresses the body's natural production of the hormone, sometimes creating fatally low levels if the drug is suddenly withdrawn.

After the drawbacks of cortisone were discovered, arthritis researchers fell back on a different group of medicines, the non-steroidal anti-inflammatory drugs, or NSAIDs. Like steroidal drugs, these medications proved useful in relieving the pain and discomfort of arthritis, and millions of people now take them every year. Many new NSAIDs have been developed in recent decades, but one of the most effective—

ACHING KNEES An X-ray shows knee joints affected by arthritis. Anti-inflammatory drugs (bottom) can ease the discomfort, but they cannot halt the disease.

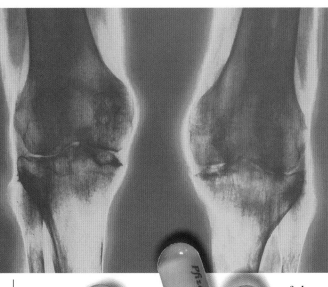

aspirin—has been in use for over a century.

After decades of research, the causes of arthritis are still a puzzle. No one yet knows what triggers the immune system to attack the joints in rheumatoid arthritis, or why normal cartilage—a slippery substance that allows bones to slide against each other—breaks down when osteoarthritis sets in. But while arthritis prevention has made little headway, some forms of treatment have been more successful. In the space of 50 years, arthroplasty—better known as joint replacement—has developed from an experimental procedure into one of the most common kinds of surgery. Until the American surgeon Marcus Smith-Petersen pioneered the first hip replacements in the 1940s, most elderly people with severe arthritis became chairbound and totally dependent on others. Today, artificial hips give renewed mobility to thousands of people every year.

Aging and the brain

In the 19th century, neurologists discovered that nerve cells, once formed, are never replaced. Skin cells or blood cells are constantly regenerated to make up for losses through wear and tear, and so are most other cells in the body, but nerve cells do not follow this rule. As a result, the number of nerve cells drops throughout adult life, so that an aging brain functions with only a proportion of the cells that it originally had. This does not necessarily mean that it functions badly. Thomas Edison was still taking out patents in the late 1920s at the age of 80, and the 20th-century British philosopher Bertrand Russell continued writing and speaking about current affairs well into his nineties. But in most people, as Ignatz Nascher noted in his textbook on geriatric medicine, the aging mind loses some of its sharpness and its ability to remember. The most prominent mental characteristic becomes an interest in one's self, rather than in the wider world.

In some cases, however, the changes are much more debilitating than this. Two disorders in particular, Parkinson's disease and Alzheimer's disease, can have a devastating

THE AGILE MIND Even at the age of 94, the philosopher Bertrand Russell took an active interest in politics and international affairs.

effect on the aging brain, but as has been shown by 20th-century research, they do so in quite different ways.

Of the two disorders, Parkinson's disease has the longest medical history, having first been described by the British surgeon James Parkinson in 1817. In his *Essay on the Shaking Palsy*, Parkinson correctly listed all the symptoms of a disease that usually strikes people over 60, and which interferes with the brain's ability to coordinate movement. Parkinson's disease usually begins with an almost imperceptible tremor of an arm or leg, which stops when the limb is used. This innocuous condition then progresses to something more serious as the muscles become stiff and slow to respond.

When he described the disease, James Parkinson had no idea what caused it. Early in this century, neurologists tracked its origin to cells in the midbrain, a 1-inch-long structure that sits at the top of the spinal cord. By the 1950s, the cells had been more precisely pinpointed to a small area of the midbrain called the substantia nigra. At about this time, researchers discovered that drugs such as reserpine could produce symptoms that seemed to mimic the disease. A key feature of these drugs was that they disrupted the production of dopamine, one of the substances

A SHIFTING FOCUS

In the first half of the 20th century, most of the major developments in medicine, such as the development of vaccines, had the greatest impact on infants, children and young adults. In the second half of the century, medical breakthroughs such as organ transplants, joint replacements and heart bypass operations had the greatest effects on people over 50. Most of these later developments correct chronic deterioration, rather than acute medical conditions.

that nerves use to pass signals to their neighbors. If Parkinson's disease did this, drugs that restored the level of dopamine might reverse the progress of the disease.

This was the reasoning behind a series of famous clinical trials that were carried out in 1961 using a new drug called levodopa, or L-dopa. For sufferers of Parkinson's disease,

STILL FIGHTING Boxing great Muhammad Ali, diagnosed with Parkinson's Disease, speaks to reporters at a New York hospital in 1984.

L-dopa was a landmark in medication. It could not cure the disease, but it did offer some hope of bringing it under control. Today, L-dopa is used in combination with several other drugs, and helps to relieve the symptoms of this otherwise intractable ailment.

Compared to Parkinson's disease, Alzheimer's disease has a much more recent history. It only came to light in 1906 when Alois Alzheimer, a German neurologist, carried out an autopsy on a 55-year-old patient who had died in a profoundly demented state. He found that the woman's brain contained deposits called plaques, which normally develop only in the very old, and he also discovered that its nerves contained tangled fibers. These tangles were easily seen under the microscope, and they became the hallmark of Alzheimer's newly identified disease.

Since 1906, the progress of this previously unlabeled affliction has been daunting. In many Western countries, it now affects 5 percent of people over 65, and is the fourth most common cause of death among the elderly after heart disease, cancer and stroke. However, on their own, these statistics tell only part of the story. Some of this steep rise is undoubtedly due to the fact that, before 1906, the disease would often have been dismissed as a form of progressive insanity. A more significant factor is the increase in life expectancy that came about with the introduction of antibiotics in the 1940s. With fewer old people prey to infections such as pneumonia, degenerative diseases such as Alzheimer's have become much more common.

Unlike Parkinson's disease, the physical changes that gradually destroy the Alzheimer sufferer's intellect have been known right from the start. But identifying changes in nerve cells is not the same as identifying their cause. In the decades since Alzheimer made his initial discovery, an

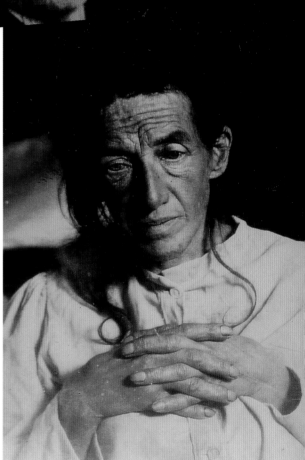

A NEW DIAGNOSIS Dr. Alois Alzheimer was the first person to describe the symptoms of Alzheimer's disease. This woman patient (below) was the disease's first recognized victim.

increasing amount of research has focused on this problem, as yet without any concrete results. Several causes have been suggested, including chronic poisoning by metals such as aluminum, and genetic factors. Since the disease appears to run in families, the genetic explanation now seems the most likely.

Theories of aging

Throughout the entire history of medicine, doctors have sought to find out why aging occurs, and to discover what—if anything—can be done to prolong life, or to slow the aging process. In some cases, this search has been driven by academic interest. In others, the prospect of fame and fortune has been an equally powerful spur.

THE MONKEY GLAND AFFAIR

A BIZARRE "REJUVENATING" OPERATION INVOLVING MONKEY TESTICLES TURNED A RUSSIAN EMIGRE SURGEON INTO AN INTERNATIONAL CELEBRITY

In the late 1890s, Sergei Voronoff, a Russian-born surgeon who spent most of his life in France, visited Cairo and carried out some rather unusual research. He investigated a number of eunuchs—men who had been castrated in childhood—and was deeply struck by their prematurely aged appearance. From this he concluded that, in men at least, the testicles must produce some chemical agent that slowed down the aging process. Without them, aging sped up.

Voronoff was not the first medical practitioner to link sex glands with aging. For several decades, doctors had experimented with the effects of testicle extracts, and some had even tried grafting animal testicles onto human patients. However, Voronoff was to pursue this form of surgery more energetically than anyone had before.

When he returned to France, Voronoff began a series of experimental grafts with animals. Then, in the 1920s, he turned his attention to humans. He decided to use monkey testicles because monkeys are physically similar to humans and because, as he put it, "the securing of human glands presents serious obstacles."

Voronoff's first attempts at grafts involving human patients ended in failure. The recipients developed severe infections, which was hardly surprising as the operation involved roughening the surface of one of their own testicles, and then applying slices of monkey testicle to it. But Voronoff managed to overcome these problems, and was soon carrying out the operation on an ever-increasing scale. His elderly male patients testified to the operation's success, and Voronoff's scientific publications reported positive effects on their "virility"—an outcome that turned a trickle of referrals into a stampede.

By the mid-1920s, Voronoff had carried out more than 1,000 rejuvenating operations, and had become a rich man. Despite a complete lack of objective evidence that the operation actually worked, he was fêted by the medical establishment, commended by the scientific press, and awarded the French Legion of Honor. He had many admirers in Europe and North America, although in

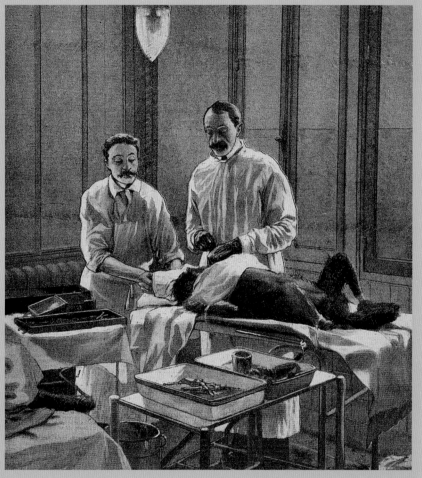

UNKIND CUT An illustration from a 1922 edition of *Le Petit Journal* shows Voronoff (also shown, left) and an assistant preparing to operate on a monkey.

Britain, a strong antivivisectionist movement prevented him from obtaining a license to carry out the operation.

In the early 1930s, Voronoff's fortunes took a downward turn. The successful isolation of testosterone—the male hormone produced by the testicles—gave researchers an easy way to replicate Voronoff's work without the need for surgery. But when they tried it, they found none of the rejuvenating effects that he had claimed. Official reports began to cast doubt on Voronoff's results, and to make matters worse, other physicians had now climbed aboard the "monkey gland" bandwagon. Although some shared Voronoff's conviction and professional expertise, others were little more than quacks.

By the mid-1930s, the monkey gland era was drawing to a close. The final blow came in the 1940s, when research into the immune system showed that "foreign" grafted tissue was quickly attacked and killed. But Voronoff stuck to his beliefs, and continued to defend his work. He died in 1951, closing an extraordinary chapter in the hunt for eternal youth.

SUPER-CENTENARIAN Jeanne Calment celebrated her 120th birthday in 1995. At the time, she was the world's oldest woman.

The Russian-French zoologist Elie Metchnikoff, who made important discoveries about the immune system, was already famous when he began his studies on aging at the beginning of the 20th century. In 1907 he published *The Prolongation of Man*, in which he developed his theory of "autointoxication" as an explanation for the aging process. Metchnikoff believed that aging was the result of waste matter that slowly accumulated in the digestive system. According to his theory, toxins from waste were absorbed by the rest of the body, and this stimulated immune-system cells to attack the body itself. As Metchnikoff had discovered phagocytosis, the process by which immune-system cells engulf their targets, his views carried some weight.

"Autointoxication" found prominent support, particularly in the person of Alexis Carrel, a French surgeon who took up work in the United States. Carrel's work on tissue culture showed that cells could be made to live an unusually long time outside the body, perhaps because they did not become "intoxicated" by its wastes. But apart from starting a vogue for drinking sour milk, which Metchnikoff recommended because its bacteria could neutralize toxins, the autointoxication theory had no practical outcome. In the scientific world at least, the idea

ACTIVE AGING Members of an over-60 club on a run. A few decades ago, such activity would have been unthinkable at this age.

that there might be a single cause of aging—and therefore a single way of stopping it—started to look less and less plausible.

This did not stop a number of so-called "rejuvenation techniques" from becoming fantastically successful until each was unmasked as useless. One of the pioneers of this mushrooming medical industry was Dr. Edouard Brown-Séquard, who in 1889 startled the French Academy of Sciences by announcing that he had regained his youthful vigor with the help of injected extracts of semen and mashed dog's testicles. Following in Brown-Séquard's footsteps, during the 1920s the Russian-born surgeon Sergei Voronoff created an immense amount of interest around the world with a "rejuvenating" operation that he performed on elderly male patients.

In the same decade, the Californian surgeon L.L. Stanley injected crushed animal testicle cells into more than 1,000 patients, and in Austria, Dr. Eugen Steinach pioneered a technique involving vasectomy in men, which again was held to turn back the clock. But no hard evidence came to light that any of these procedures had the desired effect.

Compared to rejuvenation techniques, the investigation of aging itself has been a more fruitful field of research. One of the most interesting findings came in the 1940s, when an American researcher named C.M. McKay found a link between aging and diet in laboratory rats. By feeding rats a diet that contained all the necessary vitamins and minerals, but a reduced number of calories, he found that he could extend a rat's life span by at least a third. Studies on humans—particularly members of long-lived peoples such as the Abkhazians and Hunzas of Central Asia—provided evidence that a lean diet does indeed help to prolong life.

With the discovery of DNA in the 1950s and the explosion of knowledge about the body's chemistry, research into aging has revealed processes that have echoes of Metchnikoff's theory of autointoxication. Many scientists today believe that the body's cells age because they gradually build up unstable molecules called free radicals. These can damage many essential substances, including proteins and the chemicals in cell membranes, and also DNA itself.

In healthy cells, free radicals are dealt with by a battery of special enzymes that transform them into less toxic substances, but in aging cells this chemical conversion seems to be slowly overwhelmed. Research is currently under way to devise drugs that will bolster this conversion process. If it succeeds, many of the medical problems of aging—from cataracts to Alzheimer's disease—might be preventable at last.

HEALTH AT LARGE

AFTER THE URBANIZATION OF THE 1800S, THE WORLD'S INDUSTRIAL COUNTRIES ENTERED THE 20TH CENTURY WITH A NEW WEAPON AGAINST SICKNESS: ORGANIZED MEASURES AIMED AT PROMOTING PUBLIC HEALTH. A CENTURY LATER, FOLLOWING VACCINATION AND EDUCATION CAMPAIGNS THAT HAVE SPANNED THE GLOBE, MEDICINE HAS COMPLETED THE TRANSITION FROM A MATTER THAT CONCERNS THE INDIVIDUAL TO ONE THAT CONCERNS THE WHOLE WORLD.

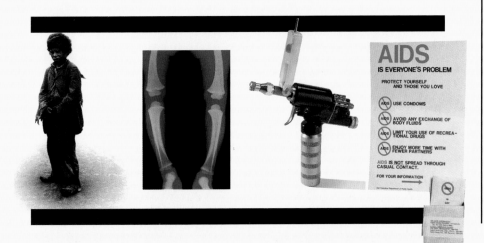

AIDS
IS EVERYONE'S PROBLEM

PROTECT YOURSELF
AND THOSE YOU LOVE

AIDS USE CONDOMS

AIDS AVOID ANY EXCHANGE OF
BODY FLUIDS

AIDS LIMIT YOUR USE OF RECREA-
TIONAL DRUGS

AIDS ENJOY MORE TIME WITH
FEWER PARTNERS

AIDS IS NOT SPREAD THROUGH
CASUAL CONTACT.

FOR YOUR INFORMATION

PUBLIC HEALTH WORLDWIDE

COORDINATED INTERNATIONAL ACTION HAS HELPED TO DIMINISH DISEASES THAT ONCE THREATENED MILLIONS AROUND THE GLOBE

Until relatively recently in human history, health was a matter for the private individual. Two hundred years ago, some civic authorities did take steps to organize basic sanitation, and quarantines were imposed from time to time to stem the spread of epidemics. But without any concrete knowledge about the cause and spread of disease, public measures such as these were often ineffective.

The turning point in public health came in the 1860s, when the French chemist Louis Pasteur set out his germ theory of disease. Once the connection between health and hygiene had been firmly established, public health measures could be put on a more scientific footing. By the beginning of the 20th century, large-scale initiatives such as sewer-building had brought about great improvements in general hygiene, and vaccination was starting to make inroads against diseases such as diphtheria. Although the discovery of antibiotics still lay four decades in the future, these measures alone triggered a sharp fall in the number of deaths from infectious diseases.

In many countries in northern Europe, this drop in the death rate reversed a rise that had occurred during the early days of industrialization and helped to spark a massive wave of population growth. At a time of expanding trade and accelerating communications, public health measures began to take on increasing importance.

By 1900, health was no longer something that could be left in the hands of the ordinary man and woman, or even their local doctor. Instead, health-promoting measures were becoming increasingly coordinated

LET THEM HAVE MILK An affectionate caricature of Dr. Gaston Variot, one of the pioneers of free milk distribution in France.

and regulated by the authorities, first within individual countries and then around the world as a whole.

Child health

In the early years of the century in America and Europe, there was widespread anxiety about the general health of the young. Some of this was generated by humane concern, but there was also a feeling—particularly prevalent during times of military draft—

DAILY DOSE German schoolchildren in the early 1920s reluctantly line up for their daily spoonful of cod-liver oil.

1900

1907 First
international health
agency founded

1911 Research
into the causes
of rickets

1918-9
Influenza
pandemic

1920
Vitamin D
isolated

1943 The drug
LSD is
discovered

1948 WHO established
Launch of Britain's
National Health Service

1950

THE NEEDY AGE Two French children—one barefoot—at the beginning of the century. In 1900, malnutrition was as much a health threat as infectious disease.

quickly became apparent, and Variot's example was copied in other countries. By the early 1900s, Nathan Straus, a wealthy American philanthropist in New York, was providing the children of New York with more than 3 million free bottles of milk a year. Like Variot, Straus used milk as a way of introducing mothers to the principles of child health: his special "milk stations" distributed information about child care as well as milk itself.

By the end of the First World War, schools came to play an important part in fostering the health of the young. While free milk helped infants and toddlers, free school meals were aimed at older children. Again, France was one of the first countries to adopt this simple but effective measure, and other countries then followed suit. The results could be startling. In a study in the English industrial city of

Bradford, children given school meals gained an average of 2½ pounds during a three-month period, while those who went without gained less than half that amount. This disparity marked the difference between healthy development and borderline malnutrition.

Once at school, children's health could be monitored much more easily than in the home. Compulsory health inspections began in France in the early 1880s, but they were at their most thorough in the United States two decades later, where

that sickly children would never grow up to help nations defend themselves. These fears were based on solid foundations. During the Boer War of 1899-1902, British medical officers had to reject three men out of every five potential recruits, many of whom had just left school. In 1917, when the United States entered the First World War, over 50 percent of drafted men were found to have physical defects. Some of these defects were the result of infectious disease, but many were caused by an inadequate diet.

At a time when the poorest people in society were often malnourished, the provision of sufficient food to infants and children became a priority in campaigns to improve public health. In 1893 the French doctor Gaston Variot showed what could be done when he opened a clinic in a run-down part of Paris, where sterilized milk was given out free of charge to the children of needy mothers. The health benefits of this free milk

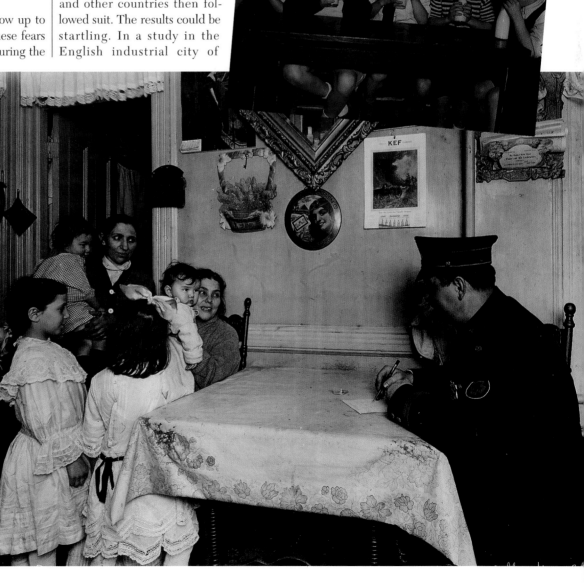

UNDER WATCHFUL EYES A New York health official visits a family in 1916, while British children (above) drink their daily milk.

1960s Global antimalarial program

1966 Global eradication of smallpox begins

1978 Final victims of smallpox

1981 AIDS first identified

WORTH THE WAIT After a smallpox outbreak, thousands of New Yorkers line up for free vaccinations in the Bronx in April 1947.

most school classes contained a large number of recently arrived immigrants, often in poor physical shape. At the beginning of the century, American health inspectors found that four-fifths of schoolchildren had head lice, one-fifth trachoma (an infectious disease of the eye), and many had skin infections such as ringworm and impetigo. Although these school inspections were sometimes seen as a humiliation by parents, they did ensure that childhood diseases were not simply allowed to run their course.

In the late 1930s, schoolchildren in North America and Europe still ran the risk of diseases such as polio, but vastly improved knowledge about the value of everyday hygiene and a well-balanced diet had made their mark. Compared to their parents, members of this generation had a much better start during the crucial early years of life.

Wherever they were applied, child-health measures produced valuable results. So did many public education campaigns aimed at adults, such as the drives to warn people about the dangers of tuberculosis and sexually transmitted diseases. But as measures like these multiplied, they raised an important issue that had worried politicians from the moment the century began. It was clear that rapid developments in medicine could do much to improve public health: the key question was, who would foot the bill?

Public health, private health

For the last 100 years or so, the answer to this increasingly critical question has depended on national feelings about the relative importance of individual choice and public responsibility. In the United States, where Nathan Straus made his mark on child health by giving away milk, fellow philanthropists such as John D. Rockefeller later trod a parallel path, ploughing millions of dollars into medical research. But while federal and state governments initiated many public health measures, government never took on a major role in organizing health care for the individual. Instead, doctors and hospitals vied with each other in an open market, competing for patients who either paid for treatment themselves, or who were paid for by medical insurance.

The outcome many decades later is a medical system that has the most lavishly funded facilities in the world, and which gives patients an enormous range of choice. But critics of the American way point out that despite this vast pool of medical expertise, the system has an underlying weakness. Although market-driven health care can deliver the finest treatment currently available, it tends to treat patients according to their financial resources rather than their medical needs.

On the other side of the Atlantic, national health care systems evolved in a variety of ways. In many countries in western Europe, the state took on a more interventionist role, ensuring that every adult contributed to some kind of health insurance scheme. In France and Germany, these schemes were often used to pay private doctors and hospitals—as they still are today—but in Britain, developments followed a different path. There, national insurance subscriptions were used to create the ultimate in state-administered health: a free health service designed to cover every person in the land.

The foundations of Britain's National Health Service were laid in 1911, when an obligatory insurance scheme was set up for working men. But the real impetus behind it came with the Second World War, when an emergency medical service was set up to

PROPOSING REFORM President Clinton speaks to Congress about health care reform in 1993. Health care was central to Clinton's 1992 campaign for the White House.

DEFICIENCY DISEASES

CAREFUL INVESTIGATION OF HUMAN AND ANIMAL DIETS LED TO THE DISCOVERY OF VITAMINS, AND TO NEW WAYS OF TACKLING SOME DEBILITATING DISEASES

A t the end of the 19th century, researchers were aware that diseases could be caused by the presence of microbes. The idea that some diseases could be caused by the absence of something was much less commonly known. The story of "deficiency diseases" dates back at least 200 years, to the discovery that citrus fruit could help combat scurvy. In the late 19th century, a Dutch doctor named Christiaan Eijkman showed that beriberi, a disease once common in the Far East, was caused by a lack of a particular substance in the diet. Many years later, this was identified as vitamin B_1. Following this lead, the British researcher Edward Mellanby began a study in 1911 to unearth the cause of rickets, a childhood bone disease that was common in northern England's industrial towns.

At the time, many experts believed that bacteria was responsible for rickets, while others suggested that it was caused by poisons in food. Mellanby pursued the food connection, setting out to see if he could devise special diets that would bring about rickets in animals. He managed to do this in dogs, and discovered that some factor present in fatty food was responsible for providing protection against the disease.

This factor turned out to be present in a traditional remedy for rickets—cod-liver oil—but at that stage no one could identify it.

Research carried out in Austria after the First World War showed that cod-liver oil helped prevent rickets throughout the year, but that during summer some of the symptoms often disappeared whether children were given cod-liver oil or not. Investigations then showed that ultraviolet radiation—which forms part of sunlight—halts the abnormal bone growth that rickets can produce. In 1920, the American biochemist Elmer McCollum isolated the mystery ingredient in cod-liver oil, and it was named vitamin D. Medical scientists then found that the human body makes vitamin D when sunlight shines on the skin. A daily dose of cod-liver oil became routine for millions of children, with sunlamp therapy during the winter for those particularly at risk. By the 1940s, in the developed world at least, rickets was largely a disease of the past.

BENDING BONES An X-ray clearly shows the bowed legs characteristic of childhood rickets. Sunlamp therapy, seen here in an American clinic in the 1920s, helped to prevent the disease.

REHYDRATION THERAPY

One of the most important developments in 20th-century health care was also one of the simplest—it consisted of a premixed pack of salts and sugars that can be used to treat dehydration. During epidemics of cholera and infant diarrhea, these packs saved millions of lives, and are still used today. Designed to be added to a set volume of boiled water, each pack produces a correctly balanced drink that helps to make up for body fluid loss.

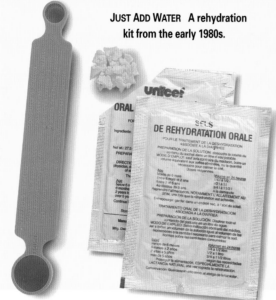

JUST ADD WATER   A rehydration kit from the early 1980s.

bring together most of the country's major hospitals and many of its doctors. This centralized system proved remarkably efficient, and when a new Labor government came into power after the war, work began on extending the service to include the medical infrastructure of the entire country. Despite opposition from some traditionalist doctors, the NHS—as it has been known ever since—was born in July 1948. For the first time, most patients left doctors' offices without paying—an experience that filled some doctors with misgivings, and patients with a degree of amazement.

Despite their different strengths and weaknesses, these 20th-century health-care systems suffer from the same fundamental problem—the difficulty of matching limited funds with constantly rising expectations. In 1900, the main challenge in public health lay in educating the population and giving them better access to basic care. Half a century later, most patients were vastly better informed about health matters, and were much more likely to consult their doctors if they felt unwell. By 1975, new medical techniques such as open-heart surgery, organ

transplants and hip replacements gave people the chance to live longer and more active lives, if they opted to have treatment. But with each new development, the cost of health care continues to spiral upward, a trend that shows no signs of coming to a halt.

Global action against disease

In the early 1900s, when organized public health was still relatively new, medical care was already beginning to take on a global dimension. The European powers had introduced Western-style medicine to many of their colonies, and a wave of research was under way into tropical diseases. At first, this research was carried out mainly for the benefit of expatriates rather than local people. As trade and travel grew, it became clear that cooperation would be needed to prevent potentially dangerous diseases from spreading back to Europe.

The first international health agency was established in Paris in 1907. Its role was largely advisory, and lay in passing on information about epidemics, and establishing detailed quarantine regulations for international travel. By this time, some progress had been made in understanding tropical diseases. Malaria and yellow fever, for example, had both been shown to be spread by mosquitoes; so had filariasis or "elephantiasis," a parasitic infection that can produce swellings in the limbs. But despite these advances in knowledge, few tropical diseases could be controlled.

After the turmoil of the First World War, and the formation of the League of Nations in 1920, the original international health agency was replaced by a permanent Health Organization in 1923. However, the League of Nations was an ineffectual body. It included only a minority of the world's countries, and its proceedings were marred by disputes. It was only after the Second World War, with the birth of the United Nations, that a truly global health structure—the World Health Organization—was formed.

From its inception in 1948, the WHO had a sizable charge. It was officially committed to "the attainment by all peoples of the highest possible level of health," and its charter defined health as not just an absence of disease, but a state of complete physical, mental and social well-being. In the decades that followed, the World Health Organization set about striving toward these ambitious aims. From the 1950s on, several WHO programs achieved major successes, while a few ended in failure.

Malaria was one of the most important targets in the WHO's fight against disease. Malaria is caused by a microscopic blood-borne parasite, and is spread by several species of mosquito that lay their eggs in stagnant water and need constant warmth in order to survive. The link between mosquitoes and malaria was proved in 1897 by Ronald Ross, and by 1912 the distinguished French bacteriologist Emile Roux predicted that this discovery would enable lands for-

BUGBUSTERS   With DDT sprays at the ready, a 1950s malaria control team arrives at a mud house in Upper Volta (now Burkina Faso).

merly forbidden to Europeans to be "opened up for civilization." But his confidence was misplaced: malaria remained a killer.

In the mid-1960s, when the WHO launched a global anti-malaria program, drugs were available to kill malaria parasites once they had entered the blood, but there was no fully effective preventive treatment. Rather than relying on drugs, as French and German programs had done in earlier years, WHO officials decided to concentrate on the disease's weak point—its insect carrier. In the early 1900s, yellow fever had been tackled by draining mosquito-ridden swamps, but the same measure achieved only limited success with malaria. By the 1950s, health workers had a new and extremely potent weapon—the insecticide DDT (it was later found to be highly toxic and banned in most countries). In 1965, teams armed with DDT went into action throughout the tropics, spraying swamps, ponds and drainage ditches in an effort to kill adult mosquitoes before they had a chance to breed.

With so many health workers in the field, it looked as if malaria might at last be eradicated, or at least brought under control, but problems soon appeared. Malaria mosquitoes can breed in the tiniest pools of water, which made it impossible to spray all their breeding sites. The mosquitoes also started to develop resistance to the insecticide, so that even sprayed areas were not completely safe. To make matters worse, the malaria parasite itself continued to evolve resistance to many of the antimalarial drugs then in use, making it harder to prevent infection and to treat it once it had begun. The worldwide spraying program faltered, and the disease began to reassert its grip.

Today, malaria remains one of the world's most pressing health problems, with more than 1 million people dying of the disease every year. New insecticides and antimalarial drugs are only just keeping the disease in check, and hopes of control are focused on developing a vaccine. But because the malaria parasite has such a complex life cycle, a successful vaccine is still some way off.

Smallpox, another major disease targeted in the 1960s and 1970s, presented the WHO with a quite different set of problems. In the mid 1960s, smallpox killed about 2 million

QUICK WORK A vaccinator at work in Bangladesh. The inoculation gun (left), introduced in the 1980s, uses air pressure to fire vaccines into the skin.

alien form of treatment. But persistence paid off, and by the early 1970s the Americas and the Middle East were largely free of the disease, although Africa and the Indian subcontinent remained a problem.

As the net closed in, politics and public unrest created immense complications, particularly when civil war broke out in Pakistan in 1971. The country had been

PRECARIOUS LIFE Her body covered in blisters, a young smallpox victim from the Congo forlornly faces the camera in 1963.

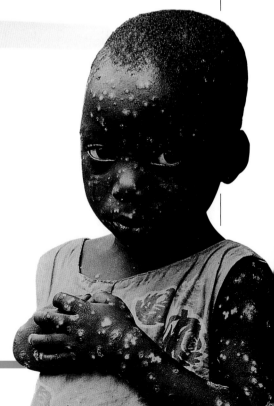

people every year, despite the fact that an effective vaccine had eradicated the disease from most of the developed world. But smallpox did have some features that made it simpler to tackle than malaria. The smallpox virus could only survive in the human body, which meant that there were no animal carriers to worry about. Also, compared to the malaria parasite, the virus produced instantly recognizable skin blisters, making outbreaks easy to identify. If enough people were vaccinated in areas at risk, the virus could—in theory—be completely wiped out.

After several false starts, the global eradication program got under way in 1966. Teams of vaccinators set off into the areas where smallpox was still endemic, and often faced a difficult task persuading people to undergo an

freed from smallpox, but with millions of refugees on the move, the disease broke out once again. The situation soon became critical, and thousands of extra workers were called in to help with the task of identifying outbreaks. In Bangladesh—the new nation that had previously been East Pakistan—nine-tenths of the country's households were inspected by health inspectors in just over a week, and everyone in outbreak areas was vaccinated. Then, in October 1974, just as the vaccination program seemed to be on the verge of success, disaster struck when the River Brahmaputra flooded. The floods triggered another wave of refugees, followed by a major outbreak of the disease. Concerted action began to win back lost ground, and by mid 1975 Bangladesh—which had become smallpox's last stronghold—was the scene of the closing chapter in the fight against the disease.

The last victim of *Variola major*, the most dangerous form of smallpox, was a three-year-old Indian girl who caught the disease in October 1975. Two years later, WHO health workers in a remote part of Somalia tracked down the last recorded case of *Variola minor*, a less dangerous form of the

THE FINAL VICTIMS OF SMALLPOX

On December 9, 1979, a group of officials at the World Health Organization's headquarters in Geneva put their signatures to a historic parchment. In half a dozen languages, it announced that smallpox had been eradicated worldwide. The signing of the document would have come earlier but for an alarming event: an outbreak of smallpox in an English laboratory ten months after the disease had apparently been conquered.

On August 11, 1978, Janet Parker, a medical photographer at the University of Birmingham, England, suddenly fell ill with a fever. Four days later she developed a rash. By August 24, Mrs. Parker's skin was covered by fluid-filled blisters and doctors began to suspect that smallpox might be involved. The disease was diagnosed on August 27, but despite efforts to save her, Janet Parker died about two weeks later.

As soon as the disease was identified, an emergency health team was set up to vaccinate everyone who had been in contact with the patient during her illness. On September 1, Janet Parker's father developed a fever, and died from a heart attack a few days later. On the 8th, her mother also fell ill and developed a smallpox rash, but eventually recovered. The days ticked by, but no further cases came to light. The source of the infection was not hard to guess. Janet Parker

FATAL ENCOUNTER Mystery still surrounds Janet Parker's accidental infection by the smallpox virus.

had worked in a building that also housed a laboratory where smallpox virus samples were held. No one who worked in the smallpox laboratory itself was affected, but Janet Parker's studio and darkroom were on the floor above the lab. Despite a detailed official inquiry, the route by which the virus reached her was never established.

The Birmingham outbreak vividly highlighted the dangers of keeping smallpox virus samples. Since the 1970s, there has been debate about the wisdom of this practice, but despite the hazards the WHO has decided to continue to sanction it to ensure that smallpox research can—if necessary—still go ahead. Today, there are only two sets of virus stocks, in the U.S. and Russia. Both are kept under armed guard.

disease. With this, smallpox was finally eradicated. But despite the celebrations that followed, the Somali case was not actually the last appearance of the disease. By a strange twist of fate, this occurred not in some remote part of Africa or Asia, but at a laboratory in England that stored samples of the virus. In 1978 a research photographer working in the same building contracted the disease and died. It was a stark reminder of the dangers of a deadly illness.

Studying the spread of disease

One of the reasons for the success of the smallpox program was that it was well-targeted. Detailed research into the way the disease was triggered and spread had shown that, despite its fearsome reputation, small-

LIFESAVING SCAR In Nepal, a Tibetan woman proudly displays a permanent reminder of her smallpox vaccination—a tiny circular scar.

pox was not quite as contagious as was often supposed. If the virus could be restricted to small enough areas, where few people were exposed to it, the disease would die out like a fire starved of fuel. This conclusion came through the scientific study of the causes and spread of disease in populations, known as epidemiology—a powerful weapon in improving public health.

The science of epidemiology dates back to the year 1854, when John Snow, a British doctor, deduced that cholera was transmitted by contaminated water. During a cholera outbreak in London, he questioned victims of the disease and found that they had all drunk water that originally came from the same pump. When he had the pump taken out of action, the number of cholera cases fell, confirming his conjecture. Snow also discovered a connection between the disease and its victims' social class. Cholera was more common among the poor than it was among the rich. This was because rich people

people were admitted to the hospital, hoping to find something that set hospital patients apart. He concluded that the decisive moment came when patients were parted

CLEAN LIVING After Dr. Charles Nicolle (left) discovered the connection between body lice (bottom) and epidemic typhus, clean clothing (center) became essential to the fight against disease.

tended to live farther upstream along the River Thames, where the water was cleaner.

By the early years of the 20th century, public health officials were beginning to follow in Snow's footsteps. Experience with diseases such as malaria and influenza, which flared up catastrophically in 1918, showed that it was not enough simply to be able to identify the agent that caused a disease. In order to keep diseases in check, it was also necessary to know about a disease's "ecology"—how each one was spread and how it was influenced by local conditions, including climate and variations in human behavior. The new science of epidemiology applied not only to diseases that were infectious. It could also be used to study illnesses that were triggered by environmental factors.

One of the earliest breakthroughs in 20th-century epidemiology was achieved in the struggle against epidemic typhus, a disease that often flares up during war or after natural disasters. In the early 1900s Charles Nicolle, a French bacteriologist working in Tunis, the capital of Tunisia, noticed that typhus victims frequently infected other people when they were at large, but rarely did so after they had been admitted to the hospital. Using an approach that has now become a standard part of epidemiology, he watched the exact sequence of events as

from their clothes, because when clothes were removed, blood-sucking body lice often went with them.

Unlike their counterparts which live in human hair, body lice do not actually live on the body itself. Instead, they inhabit people's clothes and make short excursions onto the skin in order to feed. Nicolle found that body lice are infectious only after they have fed, when they spread the disease through their droppings. When patients' clothes were taken away and washed, the parasites and their hosts were separated, and the cycle of infection was broken.

Nicolle's discovery, which was made in 1909, had important and life-saving implications. A few years later, after the outbreak of the First World War in Europe, typhus was an ever-present danger among soldiers and displaced civilians. On the Western Front, mobile laundries helped keep the illness in check, but farther east the same disease ran riot. It is estimated to have killed 3 million

people in Russia and Poland alone. Today, the science of epidemiology is at the forefront of international public health, thanks to a network of organizations such as the Centers for Disease Control in Atlanta, the Pasteur Institute in Paris and the Medical Research Council in Great Britain. By issuing bulletins on notifiable diseases—ones that have to be reported to the authorities when they occur—these organizations keep public health officials informed of changing patterns of infection. They also help the authorities prescribe preventive measures when an epidemic threatens to break out.

Before the Second World War, "germs" were seen as the main threat to public health. But with the development of antibiotics in the late 1940s and the decline of bacterial diseases, the spotlight began to

shift. During the past 50 years or so, public health measures have been increasingly preoccupied with environmental factors—that is, non-living threats to human health.

Health and the environment

Concern about environmental health hazards first came to prominence in 1945, when the release of two atomic bombs on Japan demonstrated the horrific effects of nuclear radiation. Just over a decade later, Japan was the scene of another event that made a lasting impact on fears about environmental hazards. In the late 1950s, waste mercury from an industrial plant poisoned many of the residents of Minamata, a town on the island of Kyushu. Although they were suppressed at the time, pictures of people with appalling deformities eventually found their way around the world, creating lasting anxiety about the side effects of pollution.

From the 1960s on, with revelations about the health hazards of pesticides such as DDT, of building materials such as asbestos, and of apparently innocuous substances such as artificial food colorings and sweeteners, a long public love affair with scientific and industrial progress started to come undone.

Patterns of the future

Increasingly, public health officials have had to deal not only with existing diseases, but also with the concerns of the "worried well."

THE EPIDEMIOLOGY OF AIDS

The emergence of AIDS in the closing decades of the 20th century underlined the difficulties in predictive epidemiology—the science of forecasting how fast a disease is likely to spread, and how many people it is likely to infect. During the early 1980s, when AIDS was first identified in the United States, it seemed that it would remain largely confined to homosexual men. But when infection was identified in women, and heterosexual transmission entered the equation, the original predictions were revised. It seemed by the late 1980s that the disease could affect anyone who was sexually active—a far greater number than originally thought.

These revised predictions led to a wave of public health campaigns. In Britain, for example, television commercials made it clear that almost everyone was at risk from the new disease. But as the years passed, the expected heterosexual epidemic failed to materialize, and the number of AIDS cases was lower than some early estimates had predicted. In other parts of the world, however—particularly Africa and southern Asia—differences in behavior meant that the virus struck both men and women, and triggered a wider epidemic. Infection rates were greater, and the figures are continuing to climb. In the 20 years since it first came to light, AIDS has shown how hard it is to anticipate a disease's progress, and how much human variability can influence its spread.

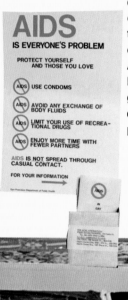

EVERYONE'S PROBLEM This 1983 sign (left) warned patrons of a San Francisco bar about the methods of contracting AIDS; below, two visitors embrace at the National AIDS Memorial quilt, on display in Denver in 1988.

This unprecedented public participation in environmental health matters shows no signs of abating. Throughout the industrialized world, the power of pressure groups has increased. Many have taken up specific health concerns, demanding rapid action. In 1900, the medical establishment led the way in drawing attention to the causes of poor health. Today, things are often the other way around, as public groups present doctors and politicians with evidence of health hazards.

INDUSTRIAL POISONING Relatives of those who died as a result of mercury poisoning at Minamata demonstrate outside a court in Japan. At the trial, the Chisso corporation was found to be responsible for the poisoning.

Sometimes both sides have been in agreement, but in many cases public perceptions and medical knowledge have been at odds.

A typical example concerns clusters of leukemia cases that have been identified around nuclear power facilities. To a lay person, the evidence looks impressive, and the connection between nuclear power and leukemia seems beyond doubt. But to epidemiologists, the matter is not so clear-cut. Many diseases occur in random clusters, and the connection between leukemia and radiation leaks is not at all easy to prove.

A similar but much more far-reaching example involves overhead electricity lines, whose powerful electric fields have been blamed for causing cancer. Some of the evidence seems persuasive, but when studied in detail the effect is so slight that it is impossible to prove.

To our forebears in 1900, when public health measures were only just starting to have an effect, such concerns might well have seemed trivial. But today, as in the past, public health still depends on constant vigilance, and on careful investigation of the patterns of disease.

A "MENACE MORE DANGEROUS THAN WAR"

DRUG-TAKING HAS A HISTORY AS OLD AS MEDICINE ITSELF, BUT IN THE 20TH CENTURY IT MUSHROOMED INTO A GLOBAL EPIDEMIC

SHOT IN THE ARM With another addict slumped behind him, an American injects himself in San Francisco's Chinatown in 1915.

that persisted into the 1920s. Drug-takers tended to be past the first flush of youth, and included many apparently "respectable" professionals, such as doctors and surgeons, who had become addicted through their work.

By the 1940s, drug-taking started to become more popular among younger people. The development of psychotropic drugs, which were originally intended to help psychiatric patients, generated a new range of mood-altering substances that could be taken for fun. In 1943, a completely new "recreational" drug was discovered when the Swiss chemist Albert Hoffmann accidentally absorbed lysergic acid diethylamide through his skin at work, and found that it was a powerful hallucinogen. Commonly known as LSD, the drug became the first of several hallucinogens to find their way into non-medical use. Like cannabis (marijuana), LSD was closely linked with the 1960s "flower power" era, when drug-taking was seen by many as a harmless and even creative activity.

Drug-taking is still on the increase among young people, and public health officials continue to search for ways to prevent it from causing harm. There is growing evidence that in the case of some drugs the crime associated with drug-taking can pose a more serious threat to health than drugs themselves.

DRUG DELIVERY Paper tabs laced with LSD (left) seem a world away from the addict's syringes and needles (below).

In November 1924, a *New York Times* headline spelled out some statistics that shocked readers. It reported that 1 million Americans were victims of a "drug habit," and that in the nation as a whole, the quantity of drugs taken illegally every year was four times the amount in Europe. The alarming rise in the number of addicts was described as a "menace more dangerous than war."

As the *New York Times* report shows, non-medical drug use is not a late 20th-century phenomenon. But during the course of the century, the nature of drug-taking changed. In the early 1900s, the principal drugs used for non-medical reasons (apart from tobacco and alcohol) were natural opiates and their derivatives—substances such as laudanum or tincture of opium, morphine and cocaine. Many of these drugs could be bought freely over the counter—a situation

MEDICINE AND THE FUTURE

THE POTENTIAL OF MEDICAL TREATMENT MAY BE ALMOST UNLIMITED—IF IT CAN BE MATCHED BY THE RESOURCES NEEDED TO FUND IT

At the outset of the 20th century, no one could have guessed that the chance discovery of a patch of bluish mold would lead to the development of the most valuable drug ever known. Similarly, no one imagined that scientists would eventually be able to identify dozens of different medical conditions by testing a few drops of blood, or that surgeons would be able to peer inside a living body and repair it by remote control. Equally, few doctors in 1900 supposed that cancer would still be a major killer 100 years later, or that a completely new disease would cause a worldwide epidemic. And at a time when the medical profession enjoyed growing self-confidence, hardly anyone guessed that patients would eventually begin to lose faith in modern medical practice, and that by the end of the 20th century people would be turning in increasing numbers to alternative therapies—such as acupuncture, reflexology and homeopathy—instead.

Predictions about the future are always bound up with the ideas and discoveries of the present, and this is particularly true of medicine. But if the overall shape of medical progress in the 20th century holds true in the 21st, future changes are likely to be diverse, complex and sometimes conflicting. There will be unexpected breakthroughs and gradual setbacks—many only visible with the benefit of hindsight—and new uses will be found for currently emerging technologies. The near future will also throw up its own share of ethical dilemmas—ones that may prove increasingly difficult to solve.

Designing new drugs

For most of the 20th century, the search for new drugs was a lengthy and repetitive business. For each new drug that has turned out to be both useful and safe, tens of thousands of potential drugs had to be rejected. The process of drug development has always been a gamble—one driven by hunches and educated guesswork, and one which all too often has failed to deliver.

GLOBAL COVERAGE Television monitors show an eye operation in progress. The operation is recorded and beamed by satellite to surgeons around the world.

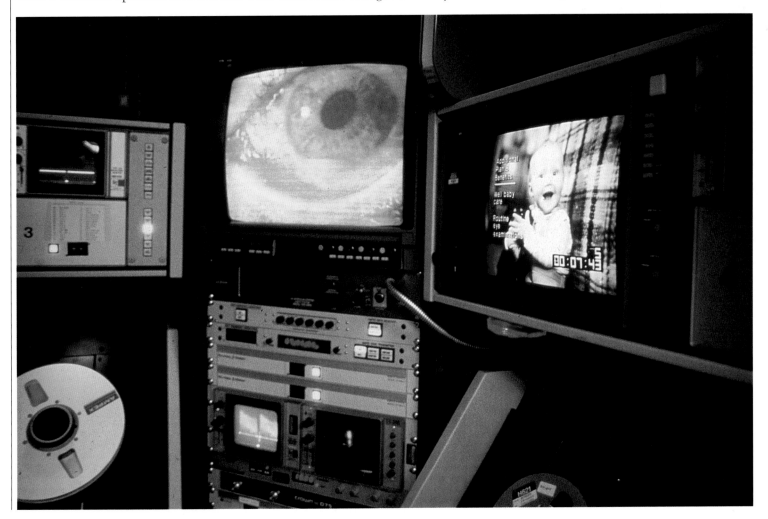

1975 César Milstein develops techniques used to make monoclonal antibodies

1976 Belgian geneticists sequence base-pairs in a virus called MS2

1977 Fred Sanger sequences øX174

1980s Human Genome Project set up

1996 UNESCO declares that the human genome is part of the "common heritage of mankind"

1997 United States grants patents for fragments of synthetic human DNA

Until relatively recently, drug researchers based their work largely on trial and error. Knowledge slowly accumulated about chemicals that had some effect on the body, but it was impossible to say exactly why most of them did what they did. The best that biochemists could do was to take a substance that seemed to have some useful property and then test similar substances that might be more potent or less toxic. This was often done by sketching a molecule's structure on a blackboard, and then guessing what changes might produce a beneficial effect.

X-RAY ANALYSIS Working with specimens smaller than a pinhead, crystallography creates X-ray patterns that reveal the structure of molecules.

This approach has one very important drawback. In the real world—including living bodies—molecules are not flat. Each kind has a distinctive three-dimensional shape. This shape provides the key to the way in which a drug works because it enables researchers to establish how the drug interacts with other substances in the body. Traditional pharmaceutical research generates molecules with different shapes, rather like a huge set of keys. The keys are then tested individually, to see if any of them turn one of the body's locks.

At one time there was no way of investigating the locks themselves in order to design chemical keys to fit them. But with the help of a process called X-ray crystallography and the immense calculating power of modern computers, this is beginning to change. Researchers can now work out the shape of many of the molecules that are found in the body, and a drug can be designed specifically to influence a particular molecule's effect.

Owing mainly to the complex mathematics involved, this form of drug development is still in its infancy, but it has some extraordinary implications. As computers become more effective, and as more is discovered about the chemistry of the human body, increasing numbers of drugs are likely to be purpose-made. Far from being a hit-or-miss business, as it was in the early years of the 20th century, tomorrow's drug research will revolve around computer-controlled design. Eventually, the result will be a pharmaceutical

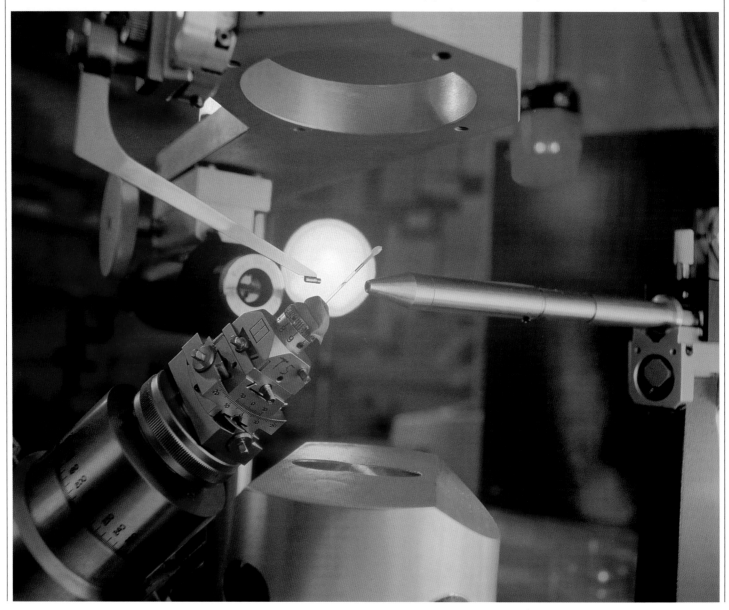

form of virtual reality, with far fewer compounds being tested but a much higher proportion producing useful results.

Monoclonal antibodies

In 1975, César Milstein, an immunologist working at Cambridge University in Great Britain, made a breakthrough that is likely to play a crucial part in the field of medicine in the 21st century.

By taking two types of cells involved in the body's immune system, he managed to combine them to form hybrid cells that had two unusual properties. First, unlike most immune-system cells, the hybrids could be cultured indefinitely in the laboratory. Second, and more importantly, the hybrids each produced large amounts of just one kind of antibody—the proteins used by the immune system to defend the body against attack. Milstein showed that it was possible to select any hybrid cell that produced a chosen

CHEMICAL KEYS César Milstein, the Nobel prizewinner who developed techniques used to make monoclonal antibodies.

HELLO DOLLY In 1997, the world's first cloned sheep— named Dolly—was revealed by a team of Scottish scientists. This development provoked many countries to ban experiments in human cloning.

antibody and make it multiply ad infinitum.

This process, called cloning, creates the biological equivalent of guided missiles. Using Milstein's techniques, biochemists can now generate "monoclonal" antibodies to lock onto a wide range of chemical targets in the body. Antibodies normally bind to target substances that have entered the body from outside, but some can bind to substances that the body itself produces. As a result, a vast number of targets— both "foreign" and "self"—can be picked out by chemical means.

In the future, monoclonal antibodies are likely to have many different uses. Because each kind of cell in the body has a distinct chemical identity, it should be possible to culture antibodies that attack tumor cells while leaving other cells completely unharmed. Monoclonal antibodies will also be exploited as delivery vehicles, carrying drugs to exactly where they are needed. Because the delivery will be so accurate, there will be far less waste, so drug doses will be cut many times over.

Monoclonal antibodies can be combined with radioactive tracers, allowing their position to be monitored from outside the body. This will enable particular kinds of cells to be pinpointed, a technique that may eventually reveal, for example, every area of a patient's body to which cancerous cells have spread. Monoclonal antibodies might even be used to seek out abnormal protein molecules circulating in the blood, putting them out of action before they have a chance to do any harm. Not only will these antibodies find the proverbial needle in a house-sized haystack, they will do the job in seconds.

Genetic medicine

During the next few decades, new drugs and antibody treatments will have a major impact in medicine, but even they seem likely to be eclipsed by probable developments in medical genetics. In 1900, the modern concept of the gene was unknown. But as the Human Genome Project nears completion, genes look set to become as important in medicine as were disease-causing bacteria a century earlier.

The Human Genome Project was set up in the mid-1980s to coordinate the work of different laboratories involved in DNA sequencing—a process that establishes the exact order of the chemical bridges, or base-pairs, that make up individual genes. When sequencing began, researchers concentrated on fragments of DNA that held genes of particular interest, but the Human Genome Project put this work on a more methodical footing. Early in the 21st century, when the project's work is accomplished, geneticists from around the world will have sequenced all the 100,000 or so human genes. In charting a total of perhaps 3 billion DNA base-pairs, they will have compiled a complete list of the chemical instructions that build and run the human body.

Once this information bank is complete, its medical value is potentially incalculable. Many diseases and disorders are known to be triggered by faulty genes—some of which come into action on their own, while others are activated by factors in the environment. If these genes can be identified, and if they can be silenced or replaced, their harmful effects may be overridden.

This kind of treatment, known as gene therapy, is still in its early stages. It has been tested against diseases such as cystic fibrosis, which can cause lasting damage to the lungs, and in this case, at least, there is already some evidence that it can have a beneficial effect. However, a gene can only work once it is inside its target cell, and it is much trick-

THE HUMAN GENOME PROJECT

BY THE EARLY YEARS OF THE 21ST CENTURY, MOLECULAR BIOLOGISTS WILL HAVE MAPPED EVERY GENE IN THE HUMAN BODY, OPENING A NEW ERA IN MEDICINE

When it is completed—in the year 2005 or thereabouts—the comprehensive map of the human genome (derived by combining "*gene*" and "chromo*some*") may turn out to be the 20th century's most important legacy to future medical science. It will represent a giant step in the study of human anatomy—a study that began centuries ago with the body as a whole, and which progressed from there to the world of tissues and cells before entering the realm of individual genes.

The human genome contains a vast amount of coded information, stored in the form of DNA. The ultimate aim of geneticists is to be able to interpret all this information, so that any damaging instructions—such as genes that cause metabolic disorders and diseases—can be identified and deactivated. Some of these genes have already been located by researchers using a technique that involves using DNA "probes." These track down genes on individual chromosomes—a process rather like tracking down individual chapters in a vast library of books. But the Human Genome Project is an altogether more ambitious undertaking, because it will eventually work through the entire library letter by letter. By the time it is complete, every one of the 3 billion letters or base-pairs will have been read, and

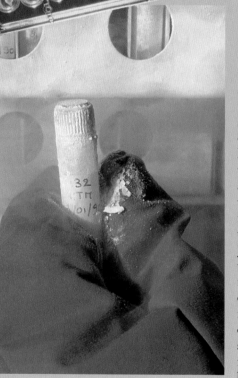

the entire sequence will be stored in electronic form.

DNA sequencing is still a relatively new procedure, but it is rapidly accelerating. In 1976, a group of Belgian geneticists succeeded in sequencing the base-pairs in a virus called MS2, but even though the virus has a tiny genome containing just a few thousand base-pairs, the work took about 15 years. At this rate, sequencing the human genome would have taken more than 500,000 years—more time than our species has existed on Earth. Just one year later, a group headed by the British geneticist Fred Sanger managed to sequence another virus, called øX174, in the space of a few months, using new techniques to cut up DNA molecules and copy them so that they could be analyzed. By the mid-1980s, automated sequencing hastened the process further—a key development in what is otherwise a tedious task.

Knowledge of the full human DNA sequence is sure to throw light on how genes work, and how they are switched on and off during normal development. It may also help to clarify one of the most astonishing revelations of modern genetics—that about 90 percent of human DNA seems to contain no instructions at all. In this century, geneticists will almost certainly discover if this so-called "junk DNA" is truly meaningless, or if it holds information that—as yet—we cannot read.

MAPPING GENES The test dishes above contain fragments of DNA—instructions for producing proteins. Frederick Sanger (left) stands by a model of insulin, one of the first proteins to have its structure fully analyzed.

ier to insert genes into living cells than to introduce a drug or an antibody into the body. In addition, to have any effect a gene must enter large numbers of cells, and it must enter them intact.

Some cells can absorb pieces of DNA, but medical researchers have found a surer way to help genes to their destination. In the case of cystic fibrosis, the genes are first copied many times. They are then inserted into viruses that normally infect the lining of the lungs but which have been treated to make them harmless. When the viruses are inhaled, they invade the lung cells, carrying their human genes with them. The genes' instructions are then put into effect, and the problem causing the disease is kept in check for as long as the virus survives.

Not all cells are as easy to reach as the ones that line the lungs, but there are alternative ways for getting genes to them. In one technique, small numbers of cells are removed from the body and given replacement genes in the laboratory. Once the cells are reinserted the genes come into action, and they keep working for as long as the cells survive.

If the technology of gene transfer can be perfected, gene therapy could help to combat a wide range of inherited diseases. At present, it is impossible to guess how far this promise is likely to be fulfilled, but some idea can be gained from the intense commercial interest it has already generated. In 1996, UNESCO—the United Nations body concerned with education and science—declared that the human genome was part of the "common heritage of mankind," and that the information within it should be treated as public property. But this idea is already running into difficulties. In 1997, the United States government decided to grant patents for fragments of synthetic human DNA that could turn out to have the same sequences as medically important genes. In the years to come, the medical use of genes is certain to increase, as are arguments about who owns the right to make and use them.

Twenty-first century technology

As well as developments in the field of genetics and biochemistry, the next few decades are bound to see new developments in medical technology. These will include new kinds of artificial body parts, new ways to deliver accurate doses of drugs at exactly

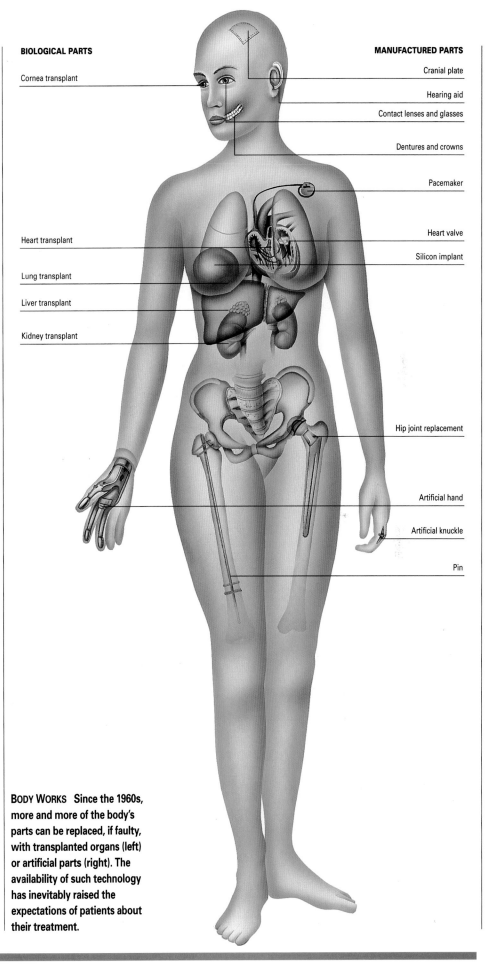

BIOLOGICAL PARTS

Cornea transplant

Heart transplant

Lung transplant

Liver transplant

Kidney transplant

MANUFACTURED PARTS

Cranial plate

Hearing aid

Contact lenses and glasses

Dentures and crowns

Pacemaker

Heart valve

Silicon implant

Hip joint replacement

Artificial hand

Artificial knuckle

Pin

BODY WORKS Since the 1960s, more and more of the body's parts can be replaced, if faulty, with transplanted organs (left) or artificial parts (right). The availability of such technology has inevitably raised the expectations of patients about their treatment.

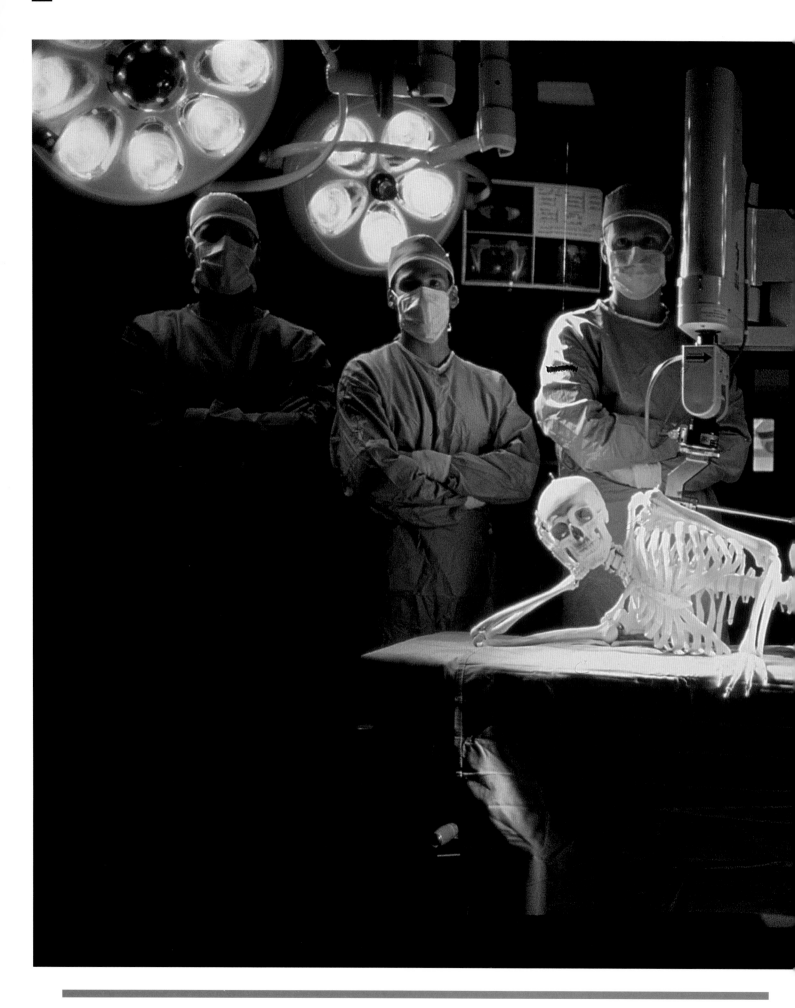

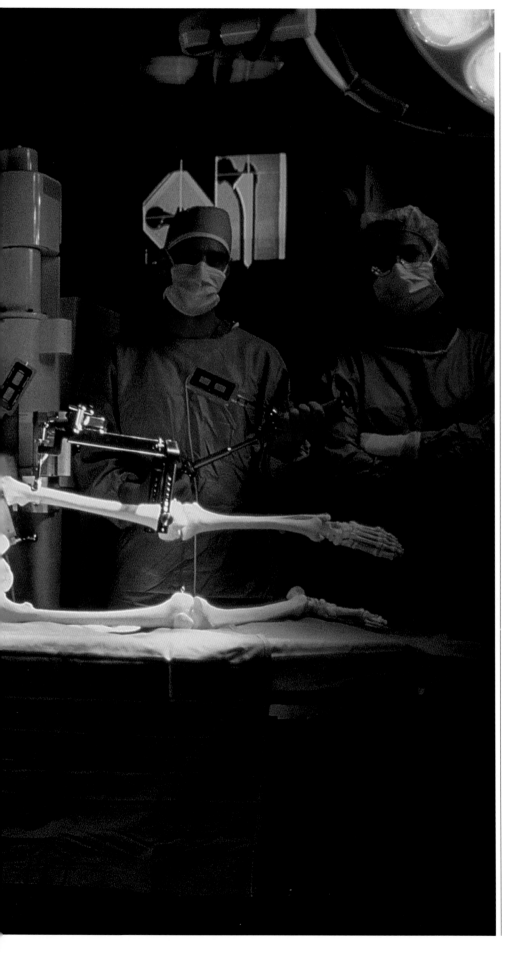

the right time, and surgical robots that can safely be entrusted with routine operations. The intrusion of robots into the operating room may sound like an alarming form of progress, but it could increase surgical safety. Machinery is already used to overcome the physical limitations of the surgeon's hands, and there is every likelihood that electronically guided equipment, such as lasers, drills and staplers, will play an increasing part in the surgery of the future.

At present, most forms of medical technology are designed to deal with problems that have already set in. During the next century, technology is likely to play an important part in disease prevention as well as treatment. Since the invention of the first effective clinical thermometer more than 130 years ago, doctors have come to rely on physical and biochemical data that reveal the body's state of health. At present, some of this data is still gathered by hand, but in the future most of it could be obtained by electronic means.

Within two decades, body temperature, blood pressure, pulse rate and cholesterol levels could all be monitored by a device as small and portable as a wristwatch. Intelligent scales could monitor changes in body weight, while electronic toilets could discreetly and automatically analyze the body's waste products for signs of abnormal constituents such as sugar or hemoglobin, the oxygen-carrying protein in the blood. These separate monitors would pool their collective data, allowing a processing unit to correlate the information and decide if any medical intervention was necessary.

In hospitals and doctors' offices, the increasing reliance on electronic data will have other far-reaching consequences. The information from personal monitors could then be loaded directly into patients' files. Case notes, X-rays and body scans will all be kept in digital form, and will be available at the touch of a button. If the expansion of computerization continues at its present rate—which seems increasingly likely—every citizen in the industrialized world could generate their own electronic health history, dating back not just to when they

DINOSAURS OF TOMORROW?

Today, we react with amazement at some of the medical institutions of the past—such as the fortress-like mental hospitals that once held thousands of patients. But it is quite possible that in the future people will be equally amazed at the dominant role of giant general hospitals in 20th-century health care.

Despite economies of scale, increased medical specialization has led to large hospitals not always being as efficient as they might be. In the future, however, improvements in communications may well mean that the original reasons for concentrating patients in large hospitals are no longer valid. For example, with the help of electronics, information about patients could be transmitted instantly to consultants anywhere around the globe, without the patient and the physician both having to be present in the same place. Some forms of diagnosis and treatment could also be carried out at a distance by electronic means.

The decline in large, centralized hospitals is likely to be mirrored by a growth in smaller clinics specializing in particular fields of medicine. With improved health monitoring, many problems are likely to be identified at an earlier stage than they are today, so less drastic treatment will be used to correct them. One advantage of this change, apart from its economic benefits, could be a return of the human face of medicine—something that is frequently lost in today's large hospitals.

MEDICAL IMAGES BY PHONE Connected to a CT (computed tomography) scanner, an image transfer system can send X-ray scans down a telephone line at the touch of a button. Photophones are used for consultations, to acquire second opinions, and at conferences.

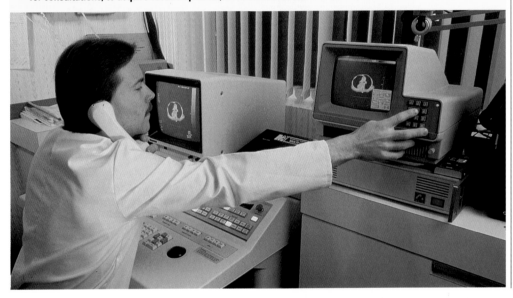

eases. Nothing could be done for anyone who lost the use of vital organs such as the lungs, heart, liver or brain. If any of them failed, the patient's life was effectively over.

Today, doctors are able to intervene in all these situations, and can often keep people alive when death would once have been inevitable. In the years to come, this capability is likely to be extended even further as new techniques are developed for detecting potential problems and replacing failing body parts. But as the last few decades have shown, this comes at a price: extremely difficult decisions have to be made about who should be treated, and how far treatment should go.

At present, techniques such as amniocentesis and chorionic villus sampling allow genetic tests to be carried out on children long before they are born. These tests can already reveal several hundred disorders, including Down's syndrome, spina bifida and hemophilia, and as more is learned about the human genome, the number is certain to grow. Knowledge about genes may eventually help in combating these disorders, but in the short term it creates some burdensome ethical dilemmas.

Some inherited disorders are so damaging that they invariably prove fatal during early development or soon after birth, but many more are less severe, although they still produce considerable suffering. In the latter

CHORIONIC VILLUS SAMPLING In the future, fetal tests may tell us almost too much about the medical prospects of children yet unborn.

were born, but to the moment when conception was first identified.

It is quite possible that this growth in electronic information will have major repercussions on the medical profession itself. With automatic diagnosis of common ailments and a system of precisely targeted drugs, much of the work that doctors carry out today could soon be taken off their hands. As technology changes, the role of doctors is bound to change as well.

Moral dilemmas

At the beginning of the 20th century, little could be done to save the lives of premature babies. There were no ways of testing for inherited abnormalities before birth, and few ways of combating most childhood dis-

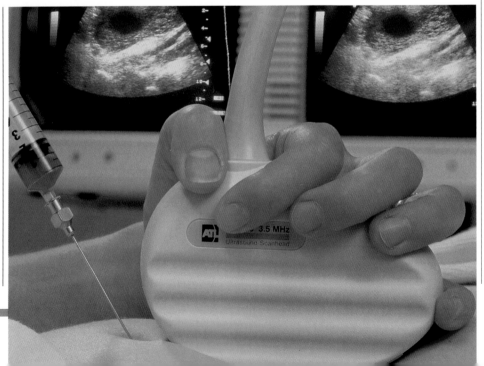

cases, the information from genetic tests could force more parents to make hard decisions about whether a new life is worth bringing into the world.

This deliberate weeding out of disability has some ominous resonances. During the early 1900s, organizations promoting eugenics—the genetic "improvement" of the human species—had a wide following among the intellectual elite in America and Europe, and for several decades a number of countries forcibly sterilized people identified as defective. But although eugenics was labeled a science, the decisions about which groups were defective were subjective. In Nazi Germany, the definition was broadened to include not only the disabled and mentally ill, but also entire racial groups. The ensuing Holocaust subsequently cast a dark shadow over the whole eugenics movement and anything that even faintly echoed its aims.

Another area of growing concern today centers on cloning. Since the 1970s, when César Milstein carried out his work on monoclonal antibodies, cloning techniques have become much more sophisticated. Biologists have discovered how to clone adult body

DOCTOR DEATH? Dr. Jack Kevorkian, left, a leading advocate of assisted suicide, grimaces after being convicted of second degree murder in Michigan in 1999. Euthanasia continues to be a controversial issue in medical ethics.

CRYONICS: THE FUTURE THAW

Recent decades have seen growing interest in cryonics—the latest in a long line of human efforts to gain immortality. In cryonic preservation, the body is drained of blood immediately after death and suffused with chilled oxygen-rich fluid. It is then preserved at low temperatures, in the hope that some future technology will allow resuscitation. However, the chances of resuscitation taking place in the 21st century look slim. Some forms of life, such as bacteria and protozoa, can survive extremely low temperatures for many years, and some human cells, such as sperm, can be stored successfully in these conditions. But so far there is no evidence that human organ systems can be revived after a protracted freeze following natural death—an essential feature if cryogenics is to work.

cells, resetting their developmental mechanisms so that they become "totipotent"— meaning that they are no longer committed to their original specialized role in the body. At present, the technique has only been carried out on laboratory animals, but the medical implications are enormous.

Cloned adult humans may still be the stuff of science fiction—and the majority of experts, together with most members of the public, believe that they should remain that way—but with cloned embryos, the issue becomes more blurred. If these are developed, they could provide tissues for transplantation, and some researchers feel that it is only a matter of time before this possibility is exploited. The technology is already in place—all that remains is for someone to put it into practice.

With each new advance in medical science, ethical dilemmas like these are certain to multiply, particularly where highly interventionist treatment is involved. Even without the additional complication of human cloning, the dilemmas concerning organ transplantation are already with us. Unless there is a great upsurge in the use of animal organs—an issue that is itself an ethical minefield—transplantable organs will remain in short supply until artificial substitutes can be perfected. Doctors will therefore continue to face difficult decisions as to which patients should be treated first.

At present, medical considerations shape these priorities, but in the future other factors may come into play. Some doctors foresee a situation in which "deserving" patients—those who have taken care of their health—take precedence over "undeserving" ones, whose unhealthy habits have led to disease. But this distinction is not as clear-cut as it seems. Although human behavior is largely shaped by learning, it is also partly pro-

grammed by genes. If certain kinds of unhealthy behavior turn out to have a genetic basis, it will be much harder to dismiss them as the patient's fault and to put these people at the bottom of the list.

At the end of life, as at the beginning, ethical dilemmas look certain to increase. In the 21st century, a larger proportion of the world's population will live to old age than ever before, and unless some sensational breakthrough is made in the science of aging, unprecedented numbers will join the ranks of the chronically ill. In the coming century of senior citizens, the debates about euthanasia, and about patients' rights to determine their own treatment, are likely to gain even more momentum than they have at the moment.

During the 20th century, medicine made extraordinary strides in abolishing childhood diseases and making adult life as healthy and fulfilling as possible. But despite these developments, and despite campaigns to come against the ailments of old age, humans will still be mortal. In this new century, one of the most important contributions medicine makes may be on the subject of how life ends.

1900

The American team headed by Walter Reed discovers that **yellow fever** is transmitted by mosquitoes.

IN THE BLOOD   Walter Reed (above); Karl Landsteiner (background), pioneer of blood transfusions.

Austrian psychiatrist Sigmund Freud publishes ***The Interpretation of Dreams***. It becomes a key work in the development of psychoanalysis.

Karl Landsteiner, an Austrian physician, discovers the **ABO blood group system**. His breakthrough makes blood transfusions safe for the first time.

The first **motorized ambulances** enter regular service.

1901

The German physicist Wilhelm Roentgen is awarded one of the first-ever Nobel prizes for his discovery of **X-rays**, made in 1896.

A Norwegian surgeon resuscitates a patient whose heart stopped during an operation by opening her chest and giving her **heart massage** by hand. The patient survives.

1902

English physiologists William Bayliss and Ernest Starling discover the digestive hormone, **secretin**.

In Germany, drug-induced *Dämmerschlaf*, or **"twilight sleep,"** becomes popular as a way of overcoming the pain of childbirth.

1903

Working with jellyfish stings, Frenchman Charles Richet discovers a life-threatening allergic reaction, later named **anaphylactic shock**.

Willem Einthoven, a Dutch physiologist, develops a rudimentary **electrocardiograph** (EKG), a machine for measuring irregularities in heartbeat.

George Perthes, a German surgeon, discovers that X-rays can be used to attack tumors, paving the way for **radiotherapy**.

1904

German chemist Alfred Einhorn develops **novocaine**, the first effective local anesthetic. Unlike cocaine, it is not addictive.

French psychologists Alfred Binet and Théodore Simon develop one of the earliest **intelligence tests**.

1905

In Germany, Fritz Schaudinn and Eric Hoffman discover the spiral bacterium that causes **syphilis**, still a wide-spread disease.

In New York, Alexis Carrel perfects surgical techniques for rejoining severed blood vessels, creating the new field of **vascular surgery**.

Nikolai Korotkoff analyzes the heart sounds that can be heard through a **stethoscope**. His work allows these sounds to be used in the diagnosis of heart disease.

Following Karl Landsteiner's discovery of blood groups, one of the earliest successful **blood transfusions** is carried out, transferring blood directly from a donor.

1906

Frenchman Jules Bordet discovers the bacterium that causes **whooping cough**.

August von Wasserman develops a simple and reliable test, the **Wasserman test,** for syphilis, a disease that often stays hidden until its final stages.

Alois Alzheimer, a German doctor, examines a female patient and identifies a form of dementia that he links to physical changes in the brain. The condition becomes known as **Alzheimer's disease**.

1907

German engineer Bernard Draeger develops the **Pulmotor**, a clockwork forerunner of the medical ventilator.

The first major international **health agency** is established in Paris.

1908

American pathologist Howard Ricketts discovers a new class of disease-causing bacteria, later named **rickettsiae**.

The antibacterial drug **sulphonamide** is synthesized in Austria. Its valuable medicinal properties go unnoticed until the 1930s, when it is discovered to be the active constituent in another drug, Prontosil.

1909

In French North Africa, Charles Nicolle discovers that epidemic **typhus** is transmitted by body lice.

BIRTH CONTROL   The Dalkon shield, one of many IUD devices.

A new female contraceptive, the **intrauterine device**, or IUD, is invented.

1910

In Germany, Paul Ehrlich introduces **Salvarsan**, the first modern drug that is effective in fighting a specific disease—in this case, syphilis.

American doctor James Herrick reports his discovery of **sickle cell anemia** (sickle-cell disease) which affects mainly black people and causes damage to internal organs.

1911

In New York, Francis Peyton Rous publishes research showing that some kinds of cancer can be caused by **viruses**.

1912

The word **"vitamin"** is coined by the Polish-born biochemist, Casimir Funk. Initially, it is spelled "vitamine," because Funk believes that all vitamins contain chemical groups called amines. Research later shows that vitamins are a very varied group of substances.

Curare is used for the first time as a muscle relaxant in surgery, making it easier for surgeons to cut into the body.

1913

At Johns Hopkins University, John Jacob Abel develops the world's first **artificial kidney**. To filter the blood, he uses membranes made of collodion, a form of cellulose also used to make photographic films.

X-rays are used to diagnose breast cancer in Germany, but many decades are to pass before **mammography** is widely used for cancer screening.

1914

In the United States, Alexis Carrel carries out the world's first successful **heart operation** on a dog.

The first **indirect blood transfusion** takes place using blood that has been citrated to prevent clotting.

British biologist Henry Dale isolates **acetylcholine**, a substance later found to carry signals between nerves. Acetylcholine, and substances like it, become known as neurotransmitters.

German cell biologist Theodor Boveri correctly suggests that cancer is triggered by **chromosome** abnormalities.

Margaret Sanger publishes ***Family Limitation***, a book that advocates birth control. It leads to her prosecution for obscenity.

1915

By carrying out tests in a Mississippi jail, Joseph Goldberger finds that **pellagra** is caused by a vitamin deficiency.

Japanese researchers Katsusaburo Yamagiwa and Koichi Ichikawa show that coal tar is **carcinogenic**—it is the first carcinogen to be identified.

1916

Margaret Sanger opens America's first **birth control** clinic.

Blood for transfusion is preserved by **refrigeration** for the first time.

1917

The natural anticoagulant **heparin** is discovered, and is later used as a treatment for strokes.

Julius Wagner-Jauregg, an Austrian psychiatrist, shows that infecting patients with **malaria** can cure many of the symptoms of syphilis.

A **vaccine** is developed against the rickettsia bacteria that causes Rocky Mountain spotted fever.

Psychiatrist Carl Jung publishes *The Psychology of the Unconscious*, one of his most important works.

1918

A catastrophic **influenza epidemic** sweeps the world at the end of the First World War, eventually killing more people than the war itself.

1919

Belgian bacteriologist Jules Bordet receives a Nobel prize for his discovery of **immunity factors** in blood serum.

1920

British surgeon Harold Gillies publishes *Plastic Surgery of the Face*, a milestone in the development of reconstructive surgical techniques.

Abortion is legalized in the Soviet Union, eventually becoming a major form of birth control.

Swiss psychiatrist Hermann Rorschach devises his famous **inkblot test** as a way of probing the subconscious.

1921

Marie Stopes opens Britain's first **family planning clinic** in London.

Swedish surgeon C.O. Nylen pioneers **ear operations** carried out with the aid of a microscope.

1922

Insulin is used medicinally for the first time, saving the life of a 14-year-old boy. Animal insulin later becomes available in commercial quantities, allowing the symptoms of diabetes to be brought under control.

Vitamin D is discovered in cod-liver oil. It proves to be an effective treatment for the bone disease rickets.

1923

The **BCG tuberculosis vaccine** is developed in France by Albert Calmette and Camille Guérin. It meets resistance in other countries on safety grounds.

The **Otophone hearing aid** goes on sale in Britain.

1924

Willem Einthoven wins a Nobel prize for his development of the **electro-cardiograph** (EKG) machine.

Acetylene gas is used as

MAN AND MACHINE   Albert Calmette, hero of the struggle against TB. Background: An early device used to record blood pressure.

a general anesthetic in surgery for the first time.

1925

Robert Robinson, a British chemist,

synthesizes **morphine** in the laboratory. Synthetic morphine becomes one of the most important painkillers used in medicine.

Sergei Voronoff, a Russian-born surgeon, becomes world famous for his **"rejuvenation therapy"** for men, which involves implanting pieces of monkey testicle into human patients.

1926

The first **portable electrocardiograph** (EKG) becomes available.

Writer Paul de Kruif captures the public imagination with *The Microbe Hunters*, a book about bacteriology.

1927

Philip Drinker and Louis Shaw design the first **"iron lung."** It proves to be a lifesaver for patients whose chest muscles are paralyzed by polio.

1928

In London, Alexander Fleming discovers that **Penicillium** mold produces a powerful antibiotic, but does not pursue its medical implications.

George Papanicolou develops the **"Pap"** test for uterine cancer.

Self-adhesive bandages go on sale, making it much easier to treat minor cuts.

British researcher Frederick Griffith discovers that genetic information can be passed from one bacterium to another by chemical means. The chemical, called "transforming principle," is later identified as **DNA**.

1929

German surgeon Werner Forssman introduces a **catheter** through an arm vein and slides it into his heart, pioneering a new way of studying heart function. Years later, he wins a Nobel prize for his work, but at the time many surgeons condemn it as highly dangerous.

1930

Doubt is cast on the **safety** of the

BCG tuberculosis vaccine when 60 babies die in the German town of Lübeck shortly after being vaccinated.

1931

English bacteriologist William Elford proves that viruses are **particles** by trapping them in ultrafine filters.

Ernest Goodpasture, an American pathologist, shows that viruses needed for **vaccine production** can be grown in eggs.

Vitamin A is isolated by Swiss chemist Paul Karrer.

Alka-Seltzer goes on sale in the United States.

ALKA-SELTZER   Persuasive advertising made Alka-Seltzer a best-seller.

The male hormone **testosterone** is first isolated by the German chemist Adolph Butenandt.

1932

German chemist Gerhard Domagk discovers **prontosil**, the first of the sulpha drugs.

First **electron microscope** built by the German engineer Ernst Ruska.

1933

Vitamin C is the first vitamin to be made synthetically. It is later made generally available, marking the start of the vitamin supplements industry.

Grantly Dick-Read publishes *Natural Childbirth*, a book that challenges drug use and other forms of intervention during childbirth.

Benzopyrene, a substance found in tobacco smoke, is found to be carcinogenic.

Manfred Sakel, a Viennese psychiatrist, uses **insulin shock therapy** to treat patients with severe mental illness.

1934

Adolph Butenandt isolates the male sex hormone **androsterone**.

1935

The first **blood bank** is set up at a hospital in Chicago.

American biochemist Wendell Stanley discovers that viruses can be **crystallized**, just like non-living substances, without losing their power to produce disease.

1936

In the Soviet Union, the world's first attempt at **human organ transplantation** ends when a patient dies after receiving a new kidney.

American researcher J.J. Bittner shows that mice can transmit **cancer** in their milk, providing evidence that some cancers are caused by viruses.

1937

In South Africa, Max Theiler develops a vaccine against **yellow fever**.

The first **antihistamine** is discovered by Daniele Bovet in Paris. Antihistamines are later used for combating allergic reactions, although side effects limit their use.

A second antibacterial sulpha drug, called **sulphapyridine**, is discovered.

1938

In Italy, psychiatrist Ugo Cerletti becomes the first person to carry out **electroconvulsive therapy**, or ECT, on a schizophrenic patient. The treatment produces an immediate improvement in the patient's condition.

British surgeon Philip Wiles develops an early version of the stainless-steel **replacement hip**.

The first plastic **contact lenses** are manufactured. Although more comfortable than earlier contact lenses, which were made of glass, they still cannot be worn for more than a few hours at a time.

Vitamin E is identified.

1939

Howard Florey and Ernst Chain isolate pure penicillin from *Penicillium* mold. With the start of the Second World War, penicillin production becomes an urgent medical priority.

French-American microbiologist René-Jules Dubos discovers **tyrothricin**, an antibiotic produced by bacteria that live in soil.

In Switzerland, Paul Müller discovers that **DDT** is an extremely powerful insecticide, opening the way for its use against insect-borne diseases.

Rhesus factor is discovered in blood. Routine rhesus factor tests later help in the treatment of erythroblastosis, a blood disorder found in newborn children.

1940

Penicillin is successfully tested on mice, paving the way for its use on human patients.

Experiments using **radioactive iodine** show that iodine is used by the thyroid gland.

1941

A previously unsuspected link is established between **rubella** (German measles) and fetal abnormalities.

Russian-American microbiologist Selman Waksman coins the term **"antibiotic."** In the years immediately following, a number of new antibiotics are discovered.

1942

A new **Yellow Fever vaccine** is developed that does not require the use of human blood serum.

1943

Swiss chemist Albert Hoffmann accidentally discovers the powerful

hallucinogenic effects of **lysergic acid diethylamide**, or **LSD**, when he absorbs the drug through his skin.

1944

The world's first **"blue baby"** operation is carried out in Canada.

In the United States, Oswald Avery, Colin MacLeod and Maclyn McCarty prove that DNA carries **genetic information**.

1945

Water fluoridation is introduced in the United States to prevent tooth decay, despite protests about the ethics of enforced medication.

In The Netherlands, Willem Kolff completes his prototype **kidney machine** and successfully tests it on human patients.

1946

Felix Bloch and Edward Purcell independently observe the phenomenon of **nuclear magnetic resonance**, which is later used in MRI, a new form of medical imaging.

1947

American surgeon Claude Beck successfully resuscitates a patient by using **electrical defibrillation** for the first time.

1948

American researchers Philip Hench and Edward Kendall find that

cortisone can be used to treat rheumatoid arthritis.

The **World Health Organization** (WHO) is formed within the United Nations.

The **National Institutes of Health** are set up in the U.S.

SEX SPECIALIST Famed sex researcher Alfred Kinsey (center), in his Indiana offfices in 1953.
Background: John Gibbon, inventor of the heart-lung machine.

In Britain, doctors begin to treat patients under the new **National Health Service**, which provides treatment free at the point of delivery.

Sex researcher Alfred Kinsey publishes his taboo-breaking book, *Sexual Behavior in the Human Male*. The book, which becomes known as the "Kinsey Report," is soon a huge best-seller.

1949

The American chemist Linus Pauling shows that sickle cell anemia is caused by a gene that produces abnormal **hemoglobin**, the oxygen-carrying pigment in blood.

In Britain, Peter Medawar discovers that the **immune system** can "learn" to accept foreign cells present during early development.

Portuguese neurologist Egas Moniz wins a Nobel prize for devising the **pre-frontal lobotomy**, a surgical treatment for some forms of mental illness.

1950

German-American biochemist Konrad Bloch shows that **radioactive isotopes** can be used to investigate chemical reactions that take place in the body.

1951

American surgeon John Gibbon completes the first **heart-lung machine**. In 1953, it is successfully used to keep a patient alive during open-heart surgery.

Kidney transplants are carried out in the United States and France, but all end in failure. In France, some of the donor kidneys are taken from prisoners executed by the guillotine.

1952

British physicist Rosalind Franklin uses **X-ray diffraction** to study the structure of DNA, and concludes that its molecules must have a spiral shape. Her breakthrough provides a clue to the way in which genetic information is stored in living things.

A **polio epidemic** sweeps the U.S., killing more than 40,000 people.

LIFE SUPPORT Cased in iron lungs, two American polio victims chat using their overhead mirrors.

In Britain, Douglas Bevis develops **amniocentesis**, a method of examining a fetus's genetic characteristics while it is still in the womb.

American surgeon Charles Hufnagel carries out an operation to implant the first artificial human **heart valve**, based on a ball-and-cage design.

The first male to female **sex-change operation** is performed.

POLIO PREVENTION Jonas Salk with testing kits for the polio virus. Background: A pacemaker.

1953

A nationwide trial of the **Salk polio vaccine** is carried out in the United States, with successful results.

James Watson and Francis Crick announce that they have uncovered the **structure of DNA**, opening the way for the development of genetic medicine.

British biochemist Frederick Sanger manages to work out the chemical structure of **insulin**. It is the first time anyone has identified all the atoms in a protein molecule.

1954

The world's first successful **kidney transplant operation** is carried out in the U.S. The donor and recipient are identical twins, which side-steps the problem of tissue rejection.

Chlorpromazine (Largactil) is introduced as a treatment for people suffering from psychiatric illness.

1955

The **external defibrillator** enters use in hospitals, and later becomes a standard item of equipment aboard ambulances.

Fluoride toothpaste goes on sale. Together with water fluoridation, it turns out to have a significant impact on the problem of tooth decay.

1956

The first **contraceptive pill** undergoes clinical trials in Puerto Rico. It cannot be tested in the United States, where it is still illegal.

1957

British bacteriologist Alick Isaacs discovers **interferons**, a group of natural defensive substances produced by cells when they are attacked by viruses.

Russian-born virologist Albert Sabin introduces the **Sabin vaccine** for polio, based on weakened live viruses.

Workers in German chemical factories fall ill with **dioxin poisoning**. Over the following decades, waste that contains dioxin becomes an increasing environmental problem.

1958

The **Page-Chayes ultra-high-speed dental drill** is introduced in the United States. It is the first drill to rotate at more than 100,000 rpm.

Fully implantable **heart pacemakers** become available. Initially they are powered by batteries, but later models use a radioactive energy source.

In Britain, Ian Donald uses **ultrasound** to examine unborn children.

1959

In vitro fertilization, or **IVF**, is carried out in rabbits, paving the way for its use in humans.

Penicillin is synthesized in the laboratory.

1960

The **"pill"** becomes available in the United States as a means of contraception. In succeeding years, it is accused of triggering lax morality.

DAILY DOSE An oral contraceptive dispenser from the mid-1960s, shortly after the "pill" is launched.

1961

Two French biologists, François Jacob and Jacques-Lucien Monod, decide that some genes—**regulators**—switch other genes on and off. Their discovery later has important implications in cancer research.

The drug **levodopa**, or L-dopa, is tested on patients suffering from encephalitis lethargica, briefly rousing them from a decades-long "sleep."

Jack Lippes introduces the **"Lippes Loop,"** a plastic IUD (intrauterine device) that can be used for birth control.

1962

After a spate of birth defects in Europe and other parts of the world, the drug **thalidomide** is identified as the cause, and is rapidly withdrawn.

Lasers are used in eye surgery for the first time.

Surgeons in Massachusetts reattach the severed arm of a 12-year-old boy who had been run over by a train. It is the first time a human limb has been successfully **regrafted**.

The **silicon breast implant** is devised. Long-term studies later cast doubt on its safety.

1963

American surgeon Thomas Starzl attempts the world's first **liver transplant** on a three-year-old boy. It does not succeed. After four further unsuccessful operations, he temporarily abandons the technique.

One of the world's most successful and profitable drugs—**Valium**—comes onto the market.

1964

Home dialysis machines become available for kidney patients in the United States and Britain.

The first **lung transplant** is carried out.

1965

A **measles vaccine** is introduced. The incidence of this childhood disease quickly falls.

Soft **plastic contact lenses**, based on hydrophilic plastics (ones that absorb water), become available for the first time. Their launch in the United States is delayed until lengthy safety tests have been carried out.

1966

The French Academy of Medicine decides that **brain death**, rather than heart stoppage, should be used as a clinical definition of death.

The World Health Organization's **smallpox eradication program** begins.

Kuru, a nervous disorder connected with cannibalism, is experimentally transferred to chimpanzees. Thirty years later, a similar disorder— **Creutzfeld-Jakob disease** (CJD)—is shown to pass from animals to people.

1967

In the United States, Thomas Starzl carries out the first successful **liver transplant**.

In South Africa, Christiaan Barnard performs the world's first **heart transplant**. The recipient survives for just 18 days, but soon other surgeons join a stampede to carry out the operation themselves.

Fertility treatment with a new drug called **clomiphene** results in a spate of multiple births.

X-ray mammography is introduced as a way of detecting breast cancer, many years after the idea was first suggested.

The **coronary bypass** operation is developed in the U.S. It gradually becomes one of the most widely performed types of heart surgery.

1968

Pope Paul VI issues **Humanae Vitae**, an encyclical that outlaws all forms of artificial contraception for members of the Roman Catholic Church.

1969

In vitro fertilization (IVF) is carried out on humans for the first time.

IVF A human egg is fertilized outside the body using a glass pipette that is thinner than a hair.

American surgeon Denton Cooley carries out the first **artificial heart implant** on a human patient. The artificial heart works, but the patient dies shortly after it is replaced with a donor heart.

The first reported outbreak of **Lassa fever** occurs in West Africa. The disease later spreads to Europe and North America.

1970

A powerful new immunosuppressive drug is discovered in a mold from a soil sample collected in Norway. This new drug, called **cyclosporin A**, eventually revolutionizes transplant surgery.

The first **nuclear-powered** heart pacemakers are produced.

The renowned American biochemist and Nobel prizewinner Linus Pauling advocates taking large doses of **vitamin C** to fend off colds and improve general health. Vitamin C sales spiral upward, despite the absence of any conclusive evidence that Pauling's suggestion actually works.

1971

Diamond-bladed scalpels are used in surgery.

1972

In Britain, a new form of X-ray imaging—**computerized axial tomography (CAT)**—is used for the first time.

British surgeon John Charnley perfects the metal and plastic **hip replacement**, which becomes a standard method of treating cases of severe arthritis.

The television drama **M*A*S*H**, based in a Korean military hospital, makes its first appearance.

1973

American biochemists Stanley Cohen and Herbert Boyer manage to insert new genes into bacterial cells, heralding the beginning of **genetic engineering**.

Magnetic resonance imaging (MRI) is available for medical use.

1974

In the U.S., William Summerlin's apparently revolutionary work on tissue rejection is exposed as a **fraud**.

1975

The world's last victim of **Variola major smallpox** is found near the border of India and Bangladesh.

In Britain, César Milstein produces hybrid cells that generate **monoclonal antibodies**.

1976

Legionnaires' disease breaks out at a meeting of the American Legion in Philadelphia. Its source is tracked down to air-conditioning tanks.

Ebola disease breaks out in Zaire and Sudan.

An accident at an industrial plant at **Seveso** in Italy releases a cloud of toxic gas containing dioxins and other poisons. Hundreds of people are evacuated so that the area can be decontaminated.

A **"bionic"** artificial limb is developed in the U.S. It uses small electric motors to mimic natural movement.

1977

Doctors in the United States begin to notice an increase in cases of **Kaposi's sarcoma**, a rare form of cancer. It turns out to be the first evidence of a new disease—AIDS.

PROBING THE BODY Raymond Damadian, inventor of the NMR (later renamed the MRI) scanner. Background: A boy gets a smallpox "jab" before going abroad on vacation.

The world's last victim of **Variola minor smallpox** is identified in Somalia.

Andreas Gruentzig develops **balloon angioplasty**, a way of unclogging arteries by stretching them open with an inflatable balloon.

1978

American researcher Robert Weinberg of the Massachusetts Institute of Technology shows that **cancer** can be caused by transferring single genes from one mouse to another.

The world's first **"test-tube baby"** is born in an English hospital, after successful in vitro fertilization (IVF). Her birth creates controversy, but IVF becomes a standard treatment for helping childless couples.

Smallpox breaks out in a laboratory in Birmingham, England, where samples of the virus were held. A medical photographer dies from it.

In the United States, the use of **chloro-fluorocarbon (CFC)** aerosol propellants is banned, amid growing fears that by damaging the ozone layer, they harm human health.

1979

A year after the last recorded outbreak of the disease, the WHO declares that the world is **smallpox-free**.

1980

German engineers develop a machine that uses **sound waves** to break up kidney stones, sidestepping the need for surgery.

An experimental vaccine against **hepatitis B** is developed.

In Switzerland, **human interferon** —a natural defense against viral infection—is successfully produced by genetically engineered bacteria.

1981

Cases of a new disease—**AIDS**—are reported in Los Angeles. Further cases soon appear in other American cities.

The world's first combined **heart-lung transplant** is carried out.

1982

An **artificial heart** designed by American Robert Jarvik is implanted in a human patient. The man survives for 112 days.

In the United States, **human insulin** produced by bacteria is authorized for general use. It is the

SPARE PARTS A Jarvik heart is removed from a patient after giving temporary life support. Background: an artificial hip.

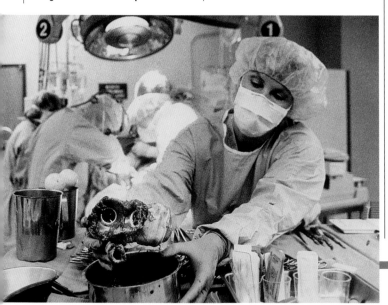

first product of genetic engineering to be licensed in this way.

The **"morning after"** contraceptive pill goes on sale.

1983

In California, developing **human embryos** are successfully transferred from one woman to another for the first time.

1984

An American baby receives a **baboon's heart** during an operation in California. Her death three weeks later sparks nationwide controversy about the ethics of using animal organs.

The first successful operation is carried out on a **fetus** in its mother's womb.

1985

Implantable **electronic defibrillators** are approved for use in the United States.

A **diagnostic test** is devised for HIV infection.

In Japan, surgeon Yuji Iwasaki transplants a pancreas and a liver from two patients who are brain-dead. Although this kind of **transplantation** is now legal in many countries, under Japanese law Iwasaki is charged with murder and corpse defilement.

A substance called **angiogenin** is discovered in tumors. By promoting the formation of new blood vessels, it allows tumors to grow.

1986

The **Human Genome Project** is set up. Geneticists begin the work of identifying the location and function of all the genes in human cells.

A gene that produces **Duchenne muscular dystrophy** is discovered.

The world's first **combined heart, lung and liver transplant** is successfully carried out in England.

Animal Liberation Front activists break into the Loma Linda Medical Center near Los Angeles, where a child had received a baboon's heart two years earlier. The break-in marks increasing hostility between medical researchers and the animal welfare movement.

An explosion occurs in a nuclear reactor at **Chernobyl**, in the former Soviet Union. In the weeks that follow,

CHERNOBYL CLINIC Young radiation victims in a leukemia ward at Kiev.

radioactive fallout spreads over the U.S.S.R. and parts of Europe.

1987

A **gene** that causes cancer of the colon is pinpointed on a human chromosome.

1988

In France, an **abortion pill** is developed containing the drug RU-486. By 2000, the drug is slowly being introduced in the United States.

1989

In France, Yvette Thibault celebrates 25 years of life with a transplanted kidney, making her one of the **longest-surviving** transplant patients in the world.

1990s

During the 1990s, genetic engineering took on an increasingly important role in medicine. With the **Human Genome Project** well under way, geneticists have already identified dozens of genes that are linked with diseases and disorders. Some of these disorders can be detected before birth. Others, such as the degenerative disease **Huntingdon's chorea**, can be identified before they start to show themselves in later life, although as yet only a few of them can be treated.

The 1990s have seen progress in surgery, particularly in the field of organ transplants. Improved immunosuppressive drugs mean that many transplant operations have a high success rate, but the number of operations carried out is limited by lack of donor organs. Research has shown that **xenotransplants**— transplants using organs from animals—could reduce the need for human donors. This possibility received a boost in 1997, with the production of the world's first cloned sheep. Carried out on animals such as pigs, cloning could produce a supply of organs for transplant surgery.

Research continues to focus on ways of preventing AIDS, and of halting the disease once it has begun. Experimental trials of **AIDS vaccines** have been proposed using modified viruses, but experts disagree about whether such trials would be safe. Meanwhile, combination therapy, which uses a mixture of different drugs, is beginning to show promising results in combating the disease itself.

Advances in medical technology have been closely linked to the phenomenal growth in the use of computers. Since the mid-1990s, **virtual reality** has been used to teach surgeons operating techniques, while computers have been used to guide lasers in microsurgery. With the rapid expansion of the Internet, the world's leading medical journals—some founded in the 19th century—are now published in electronic form.

INDEX

ACKNOWLEDGMENTS

Abbreviations
T = Top; M = Middle; B = Bottom;
R = Right; L = Left

3 Science Photo Library, L, R; Science Photo Library/Damien Lovegrove, LM; Science & Society Picture Library, RM, **6** Science & Society Picture Library, BL; Archives Department St Bartholomew's Hospital/Norman Brand, MR; **7** AKG. **8** Brown Brothers, B; Getty Images, MR. **9** From *Scientific American*, July 9, 1993, *The Great Radium Scandal* by Roger M. Macklis, TM, BR. **10** AKG, TL; Roger-Viollet/Boyer, BR. **11** Brown Brothers, ML, MR; Magnum Photos/Peter Marlow, B. **12** Siemens Medical Engineering, ML, BR. **13** Science Photo Library/Alfred Pasieka; Science & Society Picture Library, L, LM; Popperfoto, RM; Museum Boerhaave, Leiden, R. **14** Culver Pictures Inc. **15** Jean-Loup Charmet, TL; Science Photo Library/ Blair Seitz, MR. **16** Museum Boerhaave Leiden ML; Science Photo Library/ Klaus Guldbrandson, TM; Science & Society Picture Library, BR. **17** Museum Boerhaave Leiden, M, BR; Roger-Viollet/Boyer, M background. **18** Science Photo Library/Hattie Young, TL; Bilderdienst Süddeutscher Verlag, BR. **19** Science & Society Picture Library, TL; Science Photo Library/ Professor Aaron Polliak, TM; Culver Pictures Inc., BR; Science Photo Library/ Alfred Pasieka, BR background. **20** Science Photo Library, TL; Martin Woodward/Philips Analytical, TR; The Royal London Hospital Archives, BM. **21** The Royal London Hospital, BL; Ann Ronan at Image Select, BM; Science & Society Picture Library, MR. **22** Siemens Medical Engineering, TL; Graham White, MR; Jean-Loup Charmet, BR. **23** Brown Brothers, ML; Jean-Loup Charmet, MR; AKG, B. **24** From Holmes and Howry 15/Plenum Press, New York, 1963, ML; Science Photo Library, TR, MR. **25** Science Photo Library, BL, MR. **26** Ann Ronan at Image Select, TR; Siemens Medical Engineering, B; Science & Society Picture Library, B, background; small pictures L to R: Siemens Medical Engineering; Science Photo Library; Science Photo Library; Science Photo Library; Network/Rapho/Jacques Grison; Science Photo Library. **27** Siemens Medical Engineering, ML; Martin Woodward, TR; © EMI Archive, BR; Science Photo Library, BR, background. **28** Science Photo Library/Larry Mulverhill, TL; Popperfoto, MM, MR, B. **29** Topham Picturepoint, BM; Bilderdienst Süddeutscher Verlag, BR; Network/Rapho/Jacques Grison, B, background. **30** Imperial War Museum/Norman Brand, TR; Jean-Loup Charmet, MR. **30-31** London Metropolitan Archives. **31** Imperial War Museum/Norman Brand, TR; Popperfoto, M. **32** Imperial War Museum, BL; Imperial War Museum/Norman Brand, TR; Science & Society Picture Library, BR. **33** Corbis Bettmann/UPI, TR; Topham Picturepoint, BL; Frank Spooner Pictures/Gamma, BR. **34** Ullstein Bilderdienst, TL; Science Photo Library/Klaus Guldbrandson, TL, background, MR; AKG, B. **35** Roger-Viollet, TL; Bilderdienst Süddeutscher Verlag, TR; Melvin Smith/Blueprint, MR; Graham White/London Ambulance Service, B. **36** Ullstein Bilderdienst, TM; Science Photo Library, TR. **37** Image Bank/John Banagan; Science Photo Library, L, RM; Robert Opie Collection, LM; Archives Department St Bartholomew's Hospital/Norman Brand, R. **38** Science Photo Library, BL; Archives Department St Bartholomew's Hospital/Norman Brand, MR. **38-39** Roger-Viollet. **39** ©Leif Skoogfors/CORBIS. **40** Topham Picturepoint, TL; Culver Pictures Inc, TR. **40-41** Roger-Viollet/Boyer, B. **41** Archives Department St Bartholomew's Hospital/Norman Brand, TL. **42** Jean-Loup Charmet, ML. **42-43** Roger-Viollet, T. **43** Frank Spooner Pictures/Gamma, TR; Graham White, BL; From *The Construction of Modern Hospitals and their Equipment* edited by M. Schaerer S.A. Berne, Switzerland, 1935, BR. **44** ©Bettmann/CORBIS, TL; ©Bettmann/CORBIS, BR. **45** Ullstein Bilderdienst, TR; Bilderdienst Süddeutscher Verlag, TL; Culver Pictures Inc, B. **46** Roger-Viollet. **47** ©CORBIS, TR; Science & Society Picture Library, BL; Robert Opie Collection, MR. **48** Bilderdienst Süddeutscher Verlag, TR, BM; Science Photo Library, BR. **49** Science Photo Library/David Scharf, TM; Ullstein Bilderdienst, MR. **50** Brown Brothers, B. **50-51** Corbis Bettmann/UPI, T. **51** Science Photo Library/Chris Priest & Mark Clarke, TR; Brown Brothers, BR. **52** Brown Brothers, T; Getty Images, BM, BR. **53** Popperfoto, BM; Science Photo Library, MR. **54** Topham Picturepoint, TL; Science Photo Library, M; Magnum Photos/Stuart Franklin, BR. **55** Panos Pictures/Irene Slegt, TR; Tim Graham, London, B. **56** Roger-Viollet/Harlingue, ML; AKG, MR. **56-57** Roger-Viollet/Harlingue, B. **57** Science Photo Library/Alfred Pasieka, MR; Getty Images, BR. **58** Brown Brothers TM; Science & Society Picture Library, MR. **59** Getty Images, TR; Science & Society Picture Library, BL. **60** Imperial War Museum, ML, BM, BR. **60-61** Roger-Viollet, TM. **61** Science & Society Picture Library, M; Getty Images, BR. **62** Ullstein Bilderdienst, TM; Science Photo Library/Dr Karl Loumatmaa, TM, background; Science & Society Picture Library, BL. **63** Science Photo Library/Geoff Tompkinson, BR. **64** Science Photo Library, TL; Image Bank/John Banagan, M. **65** Science & Society Picture Library, TR; Roger-Viollet/Harlingue, B. **66** Jean-Loup Charmet, TL. **66-67** Brown Brothers. **68** Imperial War Museum/Norman Brand, TL; Bilderdienst Süddeutscher Verlag, BR. **69** Science Photo Library/Tim Beddow, TL; Jean-Loup Charmet, M. **70** Popperfoto. **71** Topham Picturepoint, TL; Popperfoto, BR. **72** Topham Picturepoint, TL; Science Photo Library/Ed Young, M. **73** AKG, TL; Bilderdienst Süddeutscher Verlag, MR; Corbis Bettmann/UPI, BR. **74** Ethicon Limited/Norman Brand, TL; Bilderdienst Süddeutscher Verlag, BR. **75** Science Photo Library, TL, ML, BR. **76** From *Scientific American* July, 1961 *The Artificial Kidney* by John P. Merrill, MR; Martin Woodward, B. **77** Bilderdienst Süddeutscher Verlag. **78** Science Photo Library, TL, BM; Graham White, MR. **79** Science Photo Library, TL, ML, BR. **80** Brown Brothers, TR, BL; Science & Society Picture Library, BR. **81** Popperfoto, TR; Science Photo Library/George Bernard, ML; Topham Picturepoint, BL. **82** Getty Images, ML; Martin Woodward/A & M Hearing, TR; Science & Society Picture Library, BM. **83** Science Photo Library, BL, BR. **84** Jeffrey G. Wilkinson, TL; Graham White/Nigel Gooby, B. **85** Science Photo Library/Françoise Sauze, TL, MR, BM. **86** Magnum Photos/Bruno Barbey, BL, BM; Panos Pictures/J. Holmes, MR. **87** Deborah Woodward, TL; Science Photo Library, TR; Sygma/Laurent Sylberman, BR. **88** Sygma/Jean-Jacques Grezet, TR; Corbis Bettmann, ML; Bradbury & Williams, BR. **89** Sygma/Bourdis-Dupuis, TM, BR; *Nature,* June 30, 1988, ML. **90** Magnum Photos/Marc Riboud, TR; Norman Brand, B. **91** Magnum Photos/Fred Mayer, TR, Magnum Photos/Bruno Barbey, BR. **92** Magnum Photos/Marc Riboud, TL, Science Photo Library, BR. **93** Science Photo Library, TM, TR; John Walmsley, B. **94** Science Photo Library, ML; John Walmsley, TR; Neal's Yard/Norman Brand, BM. **95** Science Photo Library; Topham Picturepoint, L; Science Photo Library, LM, R; Science & Society Picture Library, RM; **96** Roger-Viollet. **97** AKG, TL; Getty Images, BR. **98** Science & Society Picture Library, TL; ©Bettmann/CORBIS, BR. **99** Topham Picturepoint, ML; Popperfoto, TR; Science Photo Library, BR. **100** Collections/Anthea Sieveking, TL. **100-1** Brown Brothers, B. **101** Collections/Anthea Sieveking, TM. **102** Popperfoto, TM; Science Photo Library, MR; Topham Picturepoint, BL. **103** Ullstein Bilderdienst, TL; Deborah Woodward, TR; Topham Picturepoint, BM. **104** Panos Pictures/Philip Wolmuth, BL; Science & Society Picture Library, MR. **105** Bilderdienst Süddeutscher Verlag, TL; Science Photo Library, M, BR. **106** Brown Brothers, BL, MR. **107** Popperfoto, TM; Magnum Photos/James Nachtwey, TR. **108** Science & Society Picture Library, BL; ©Bettmann/CORBIS, TR; Science Photo Library, MR. **109** Culver Pictures Inc, TM; Hulton-Deutsch Collection/ CORBIS, MR; Bilderdienst Süddeutscher Verlag, BL. **110** Science Photo Library BL; Frank Spooner Pictures/Gamma/J. Chiasson, TR. **111** Science Photo Library, TR, MR. **112** Bilderdienst Süddeutscher Verlag, BM; MR. **113** Science Photo Library. **114** Science Photo Library, TR; Panos Pictures/Anders Gunnartz, BL. **115** Ullstein Bilderdienst. **116** Magnum Photos/Eve Arnold, T; Getty Images, BM. **117** Magnum Photos/Raymond Depardon, BL; AKG, TR. **118** Corbis Bettmann/UPI, TR; Graham White, B. **119** Science Photo Library, MR, MR background. **120** Science Photo Library, ML; Magnum Photos/James Nachtwey, B. **121** ©Jerry Cooke/CORBIS, BL; Science Photo Library, TR. **122** ©Bettmann/CORBIS. **123** Magnum Photos/Cartier- Bresson, BL; Sygma/Richard Smith, TR. **124** Science Photo Library, TR; Sygma/J. Pavlovsky, M; Getty Images, BL. **125** Science Photo Library, ML; Science Photo Library/Adam Hart-Davis, BL; Getty Images, TR. **126** ©Bettmann/CORBIS, BL; Science Photo Library, TM; Professor Maurer/Goethe-University, Clinic for Psychiatric and Psychotherapy, Frankfurt, MR. **127** Roger-Viollet/ Branger, BL; Jean-Loup Charmet, TR. **128** Popperfoto, TL; Topham Picturepoint, BR. **129** Science Photo Library/James King-Holmes; Roger-Viollet, L; Science Photo Library, LM; Science & Society Picture Library, RM, R. **130** Bilderdienst Süddeutscher Verlag, BL; Jean-Loup Charmet, MR. **131** Roger- Viollet, TL; Getty Images, MR; Brown Brothers, BR. **132** ©Bettmann/CORBIS, TL; ©Wally McNamee/CORBIS, BR. **133** Science Photo Library, M; Ullstein Bilderdienst, B. **134** Science & Society Picture Library, ML; Bilderdienst Süddeutscher Verlag, BR. **135** Magnum Photos/Bruno Barbey, TR; Science & Society Picture Library, M; Popperfoto, BR. **136** Popperfoto, BL; PA News, TR. **137** Jean-Loup Charmet, TL; Roger-Viollet/ND, M; Science Photo Library/David Scharf, MR. **138** ©Roger Ressmeyer/CORBIS, ML; ©Bettmann/CORBIS, BL; **138-9** Magnum Photos/Eugene Smith. **140** Brown Brothers, TL; Science Photo Library, M; Magnum Photos/ Donovan Wylie, BR. **141** Science Photo Library. **142** Science Photo Library/ James King-Holmes. **143** Popperfoto, BL; Science Photo Library, TR. **144** Science Photo Library/J.C. Revy, TR, M; Popperfoto, BL. **145** Kevin Jones Associates. **146-7** Katz Pictures/George Steinmetz. **148** Science Photo Library/Simon Fraser, ML, Science Photo Library, Saturn Stills, BR. **149** ©AFP/CORBIS, TM, TR. **150** Culver Pictures, TL; Corbis Bettmann/UPI, background; Science & Society Picture Library, MR. **151** Science & Society Picture Library, background; Jean-Loup Charmet, BM; Advertising Archives, MR. **152** Bilderdienst Süddeutscher Verlag, background; ©Bettmann/CORBIS, TR. **153** Corbis Bettmann/UPI, ML; Popperfoto, TM; Science & Society Picture Library, background; Topham Picturepoint, BR. **154** PA News, background; Science Photo Library/ Hank Morgan, TM; Bernd Auers, TR. **155** Science Photo Library/James Stevenson, background; Corbis Bettmann/UPI, BL; Panos Pictures/ Micheal J. O'Brien, MR.

Front Cover: Image Bank/John Banagan, T; AKG, M; Science Photo Library, B.

Back Cover: © Bettman/CORBIS, T; Siemens Medical Engineering, M; Science Photo Library, B.

The editors are grateful to the following individuals and publishers for their kind permission to quote passages from the books below:

Aperture in association with Many Voices Press from *Exploding into Life* by Dorothea Lynch and Eugen Richards, 1986
Blackwell Scientific Publications from *Psychosurgery in the Treatment of Mental Disorders and Intractable Pain* by Walter Freeman and James W. Watts, 1942
Journal of the American Medical Association from *Acupuncture and Anesthesia* by E. Grey Dimond, December 6, 1971
Lawrence & Wishart from *Science and Everyday Life* by J.B.S. Haldane, FRS, 1939
Mosby-Year Book Inc. from *Radiology: An Illustrated History* by Ronald L. Eisenberg, 1992
Westview Press from *Poor Women, Powerful Men: America's Great Family Planning Experiment* by Martha C. Ward, 1986